Robustness of Statistical Tests

This is a volume in
STATISTICAL MODELING AND DECISION SCIENCE

Gerald L. Lieberman and Ingram Olkin, editors
Stanford University, Stanford, California

Robustness of Statistical Tests

Takeaki Kariya

Institute of Economic Research
Hitotsubashi University
Kunitachi, Tokyo
Japan

Bimal K. Sinha

Department of Mathematics and Statistics
University of Maryland
Baltimore County Campus
Catonsville, Maryland

ACADEMIC PRESS, INC.
Harcourt Brace Jovanovich, Publishers

Boston San Diego New York
Berkeley London Sydney
Tokyo Toronto

ACADEMIC PRESS, INC.
1250 Sixth Avenue, San Diego, CA 92101

United Kingdom Edition published by
ACADEMIC PRESS INC. (LONDON) LTD.
24–28 Oval Road, London NW1 7DX

Library of Congress Cataloging-in-Publication Data

Kariya, Takeaki.
Robustness of statistical tests.
Bibliography: p.
Includes index.
1. Robust statistics. 2. Statistical hypothesis testing. I. Sinha, Bimal K., Date- . II. Title.
QA276.K226 1988 519.5′6 88–6333
ISBN 0–12–398230–8

Printed in the United States of America

88 89 90 91 9 8 7 6 5 4 3 2 1

To Shizuko, Emi, Mami
— Takeaki Kariya

To my father-in-law and the memory of my father
— Bimal K. Sinha

Contents

Preface

Much of the statistical methodology in multivariate analysis is concerned with models in which the underlying observation vectors are assumed to independently follow multivariate normal distributions. In particular, this assumption almost always prevails in multivariate nonasymptotic parameteric testing problems, with the compelling fact that the assumption preserves the linear structure of the problems and often admits transformation groups under which the problems remain invariant. No doubt, this assumption considerably simplifies the underlying analysis such as the derivations of tests, the null and nonnull distributions of test statistics, the verification of optimal properties of the tests, etc. However, in practice, the assumption of the normality of the underlying data is difficult to verify or simply may not be true. Hence, it is important to investigate the insensitivity or robustness of the procedures derived under the assumption of normality against deviations from this assumption or, equivalently, to understand the extent of distributions beyond normality under which the procedure is equally valid.

The object of this book is to develop a general, systematic finite sample theory of the robustness of tests and to apply this theory to some important testing problems commonly considered under normality. The robustness we are concerned with in this book is the exact (not approximate) robustness in which the distributional or optimal property that a test carries under a normal distribution holds exactly under a nonnormal distribution. As regards nonnormal distributions, we focus on orthogonally invariant and elliptically symmetric distributions for two reasons: (i) these distributions usually preserve the invariant structure of problems, and (ii) many distributional and optimal properties of tests that generally hold under normality are more or less associated with the invariant structure of the problems. In fact, in multiparameter hypotheses testing problems, invariance is perhaps the only successful and useful tool that allows us to consider the optimality of tests. The recent volumes by Muirhead (1982), Eaton (1983), Farrell (1985), Anderson (1984), and Giri (1977), all of which provide an excellent account of techniques leading to optimum tests (though very much with the normality assumption), bear testimony to this assertion. On the other hand, the orthogonally invariant distributions and elliptically symmetric distributions not only preserve the invariant structure of problems but also possess many similar distributional properties as the multivariate normal distribution. Furthermore, the class of these distributions is broad enough to include such heavy-tailed distributions as the multivariate Cauchy distribution, the multivariate t-distribution, contaminated normal distribution, etc. in addition to the multivariate normal distribution. In this sense, the class of orthogonally invariant and elliptically symmetric distributions is regarded as a natural class or at least as a first candidate within which the exact robustness may be studied. The present book aims at developing a unified treatment of the problem of the robustness of tests within this class.

The contents are essentially based on the work of the authors; the main tool used in the text is invariance. Elliptically symmetric distributions and their properties are discussed in Chapter 1. Chapter 2 is devoted entirely to a complete exposition of invariance. A special feature of Chapter 2 is a detailed description of what is popularly known as the representation theorem for the probability ratio of a maximal invariant due to Stein (1966), Wijsman (1967) and Andersson (1982), which is used repeatedly throughout the book. A formal proof appears in the appendix of this chapter. The basic concepts of three aspects of the robustness of tests—null, nonnull, and optimality—and a theory providing methods to establish them constitute Chapter 3, the core chapter in the book. The applications of the general theory begin in Chapter 4 with the study of the robustness of the familiar Student's t-test and tests for serial correlation. This chapter contains sections dealing with robustness without invariance, a special feature.

Chapter 5 deals at length with one of the most useful and widely applied problems in multivariate testing: GMANOVA (General Multivariate Analysis of Variance). The MANOVA, Hotelling's T^2, and ANOVA problems are discussed as special cases. Robust tests for covariance structures, such as sphericity, independence, etc., make up Chapter 6. Chapter 7 provides a detailed description of univariate and multivariate outlier problems. In Chapters 4, 5, 6, and 7 combined, readers will find the most commonly used tests under the assumption of normality.

Chapter 8 presents some new robustness results, which deal with inference in two population problems.

The book has some unique features worth mentioning. First, it provides a unified approach to a general hypothesis testing problem in a nonnormal model. Second, most standard normal-theory tests presented come out as special cases, often with simpler proofs. Third, the book brings out the central role of invariant group structure in contrast to the normal model for some problems (see GMANOVA, Chapter 5).

This book is not intended as an elementary textbook. A formal prerequisite is a graduate course in mathematical statistics. Some familiarity with multivariate tests is helpful but not required. Most of the book can be covered in two semesters. We hope that the book will stimulate further research in the hypotheses testing theory via invariance beyond normal probability models.

Acknowledgments

The first author is deeply grateful to Professors Osamu Isono, Seiji Nabeya, Clifford Hildreth, Morris L. Eaton, and Akio Kudo for their instruction and encouragement in the past. He is also grateful to his friends the Treichels and Masayoshi Mizuno for their continuing encouragement. A part of the manuscript was used by him in a course offered at the Department of Statistics at Rugers University. This opportunity is thankfully acknowledged. The second author is deeply grateful to Professors S. B. Chaudhuri, S. K. Chatterjee and H. K. Nandi, J. K. Ghosh, and P. K. Sen for their instruction and encouragement during the author's undergraduate, graduate, and postgraduate study, respectively. He is also grateful to his elder brothers for their continuing encouragement and parental care during his student days. Both authors are thankful to the late Professor P. R. Krishnaiah for his keen interest in their work and for offering financial support, and also to the University of Maryland, Baltimore County, for financial support, which made this collaboration possible. A portion of Kariya's work is financially supported by the Ministry of Education of Japan under Grant-in-Aid for General Scientific Research 61530012. Permission of the Journal of Multivariate Analysis, the Journal of Statistical Planning and Inference and Annals of Statistics to use the authors' articles in this book is thankfully acknowledged.

Finally the authors wish to express their love to the understanding wives and children for their moral support and encouragement.

Introduction

This book is devoted to a systematic development of a general theory of the robustness of tests and its application to a wide variety of multiparameter hypotheses testing problems. Robustness of a test here refers to an exact finite sample property with reference to both the null and the nonnull distributions of the test statistic concerned. The basic concepts involved in this book can be described as null robustness, nonnull robustness, and optimality robustness, which, respectively, deal with the null distribution of a test statistic, its nonnull distribution, and its optimality, if any, such as uniformly most powerful (UMP), uniformly most powerful invariant (UMPI), and locally best invariant (LBI) properties. Roughly speaking, the robustness of a test procedure here means the stability or invariance of the null distribution of the test statistic, the nonnull distribution, and the optimal properties when the underlying model undergoes change.

In the major portion of the book, the multivariate normal distribution is taken as the basic probability model, and the robustness of tests against deviations from it is the subject of our discussion. A theoretical reason for "Why normal?" can be based on the grounds that in *multiparameter* hypotheses testing problems (1) the assumption of multinormality of the underlying observations often greatly facilitates the derivation of a test procedure with some optimal properties and also the study of the null and nonnull distributions of the test statistic, and (2) replacing the multivariate normal distribution by some other distribution not only complicates the analysis of data but also, in general, leads to totally intractable procedures. The basic structure of the multivariate normal distribution that provides the property (1) is the linear or invariant structure which admits a transformation group, and optimalities of tests such as UMPI and LBI properties in multiparameter testing problems are usually associated with the invariant structure of the problems. Consequently, in this book, the invariance approach is also the basic approach to the analysis of the testing problems.

Of course, the theoretical reason cannot justify the assumption of normality, and the assumption needs to be relaxed. In Chapter 1 we discuss the point that, from the invariance viewpoint, a natural extension of the normal family is the family $\mathscr{P}_{\mathrm{E}}$ of elliptically symmetric distributions and, sometimes, the family of left-orthogonally invariant distributions, which includes $\mathscr{P}_{\mathrm{E}}$ as a subfamily. Some properties of the elliptically symmetric distributions and left-orthogonally invariant distributions are also investigated in Chapter 1. In later chapters, it will be shown that most normal-theory-based optimum invariant tests are robust against departures from normality toward elliptically symmetric or left-orthogonally invariant distributions.

The general invariance approach to testing is outlined in Chapter 2. We begin with a brief discussion of the theory of invariant measures on locally compact topological groups and then explain the notions of group actions, quotient space, and homogeneous space. These concepts are later used to describe the main result of this chapter: a representation theorem of the probability ratio of a maximal invariant due to Stein, Wijsman, and Andersson. Some conditions for the validity of the theorem are presented. The representation theorem is the main tool for establishing the nonnull robustness and optimality robustness of invariant tests.

Chapter 3 is the core chapter of this book. After briefly introducing the notions of null, nonnull, and optimality robustness of a test, most of the multiparameter testing problems commonly used in applications are listed, and the invariance structure of these problems is discussed. Then a general theory of null, nonnull, and optimality robustness is developed sequentially. At each stage, plenty of examples are presented to clarify the concepts and issues involved. These examples are repeatedly used to make clear the fine differences among the various concepts. The general theory of nonnull and optimality robustness as developed here depends heavily on the representation theorem which is derived in Chapter 2 and repeatedly used in later chapters to solve many special problems of interest. Some basic references in this chapter are Kariya (1981a), Eaton and Kariya (1984), and Kariya and Sinha (1985).

Chapter 4 is devoted to the robustness of Student's t-test and tests for serial correlation; this is the only chapter dealing mostly with robustness without invariance. The robustness of UMP and UMPI properties of the t-test and tests for serial correlation is established in many different nonnormal classes, and the various classes of tests for which the optimality properties hold therein are compared. The results are based on Kariya and Eaton (1977) and Kariya, Sinha and Giri (1987).

The general multivariate analysis of variance (GMANOVA) problem, which includes as a special case the multivariate analysis of variance (MANOVA) problem, the problem which has drawn the attention of many researchers all over the world for more than fifty years, is described in detail in Chapter 5. Many examples are presented to ensure an understanding of the subtle differences between the MANOVA problem and the GMANOVA problem. A thorough and systematic exposition on the GMANOVA problem and its extension under normality is found in Kariya (1985). The robustness properties of the tests in the MANOVA problem and in the GMANOVA problem are established in this chapter. This includes, in particular, the optimality robustness of Hotelling's T^2-test and the ANOVA F-test. The general theory developed in Chapter 3 is applied to show that for the MANOVA problem, the LBI test under normality derived by Kariya (1978) is robust for a fairly general class of elliptically symmetric distributions (Kariya and Sinha, 1985).

Chapter 6 deals with robust tests for a variety of covariance structures. Through a delicate analysis, it is shown that the standard normal-theory optimum invariant tests for independence, sphericity, and a specified covariance structure are robust for distributions far beyond normal. The result is based on Kariya and Sinha (1985).

In Chapter 7, the famous problem of the detection of outliers initiated by Ferguson (1961) and generalized by Schwager and Margolin (1982) is described in detail. Outliers in respect to both mean slippage and dispersion slippage are considered in a wider framework, and the test based on Mardia's multivariate kurtosis (Mardia, 1970) is shown to be LBI for both mean slippage and dispersion slippage. The results are based on Sinha (1984) and Das and Sinha (1986).

A different kind of approach to robustness is taken up in Chapter 8, where the independence of samples from different populations is retained. In particular, the problems involved in comparing two populations in terms of their location and scale parameters are considered. As the underlying model, we treat exponential and normal theory. It is shown that the standard normal-theory- and exponential-theory-based tests for the equality of location and scale parameters are optimality-robust. However, there is a serious lack of null robustness in all cases (Kariya and Sinha, 1987).

Chapter 1 Spherically Symmetric Distributions

1.1. Why Normal and Why Not Spherical?

In statistics, the role of the normal distribution may be viewed from the following two points:

(1) As an underlying distribution, it describes or approximates the process of data generation.
(2) As a limiting distribution, it approximates the sampling distribution of a statistic (e.g., an estimator or a test statistic) when sample size is large.

In many parametric hypothesis testing problems of finite samples, especially in multivariate testing problems, the role of the normal distribution associated with (1) is of great importance, chiefly because, as can be pointed out, the normal distribution fits the framework of the Neyman–Pearson testing theory very well. In general, a statistical procedure including a test procedure is required to be "efficient" in some sense. As is well known, the efficiency of a test procedure is evaluated as its average probability of detecting the falseness of a hypothesis when it is indeed false, which is called the power of the test (see Section 2.3 of Chapter 2). Hence a test that has as large a power as possible is desirable. Such an optimal test that maximizes its power is often provided when normal distribution is assumed as the underlying distribution. This is particularly true in a wide variety of multivariate parametric testing problems of finite samples. In fact, in the multivariate case, it is difficult to obtain such an optimal test under any other distribution. Of course, the assumption of normality is not always satisfied in real data, and the validity of the assumption must be checked before the optimal tests are applied. However, sometimes the optimal tests derived under normality are effective even under nonnormality or, in other words, the optimal tests remain robust against departures from normality. This is true for many multivariate tests. In such situations when robust optimal tests exist, the assumption of normality is considered as a means to derive the optimal tests.

On the other hand, the deviations from normality can be of a different variety; thus, the robustness of tests can be discussed in various ways. One meaningful approach may be to define a metric (distance) on the set of all distributions with certain properties and to consider the robustness of a test against the departures from normality in terms of the distance. However, optimal properties of tests are not necessarily topological properties. In other words, even if a given distribution is close to the normal distribution in the metric, an optimal test under normality is not necessarily optimal, or even approximately optimal, under the given distribution. Another approach, which is in fact our approach, is based on the following characterization of normality (see Exercise 1.1):

(1.1)

$$\mathcal{L}(x) = N(0, \sigma^2 I_n) \quad \text{if and only if} \quad \begin{cases} \text{(a)} & x_1, \ldots, x_n \text{ are independent} \\ \text{(b)} & \mathcal{L}(x) \text{ is spherically symmetric,} \end{cases}$$

where $x = (x_1, \ldots, x_n)'$ is an $n \times 1$ random vector, $\mathcal{L}(x)$ denotes the distribution of x and $N(a, A)$ denotes the normal distribution with mean vector a and covariance matrix A. Here the spherical symmetry of a distribution is defined as follows.

Definition 1.1. The distribution $\mathcal{L}(x)$ of an $n \times 1$ random vector x is said to be spherically symmetric if the distribution of x remains the same under the orthogonal transformations:

$$\mathcal{L}(\Gamma x) = \mathcal{L}(x) \quad \text{for all} \quad \Gamma \in \mathcal{O}(n), \tag{1.2}$$

where $\mathcal{O}(n)$ denotes the group of $n \times n$ orthogonal matrices.

Now because (a) and (b) in (1.1) are mutually exclusive unless $\mathcal{L}(x)$ is normal, either (a) or (b) can be assumed for nonnormal distributions. Note however that the adoption of sphericity (b) sacrifices the independence of samples. The sacrifice of independence is of course a serious problem from a practical viewpoint, because data in samples collected by a controlled experiment or data in samples randomly chosen are usually independent. On the other hand, as is discussed in Section 1.2, the adoption of independence destroys the invariant or linear structure of the normal model. This implies that optimal tests derived under normality will lose their optimalities, because most optimal properties of these tests are more or less dependent on the invariant structure. In addition, the size of a test under normality, which is the error probability of rejecting the null hypothesis when it is true, is rather sensitive to nonnormality in the framework with independence, whereas in the framework with spherical symmetry, the size of a test remains unchanged in most important problems. Therefore, in finite sample testing problems, it will be important to adopt the spherical symmetry and to study how much robust commonly used tests are against departures from normal distribution toward spherically symmetric distributions. *This is the theme of the present book.* Some practical implications of the assumption of spherical symmetry are also discussed in the next section.

Before this section is closed, we shall give a basic notation concerning normal distribution. Let $\mathscr{S}(p)$ denote the set of $p \times p$ positive-definite matrices.

Definition 1.2. Let U be an $n \times p$ random matrix and let u_i: $1 \times p$ be the ith row of U $(i = 1, \ldots, n)$. Then the notation

$$U \sim N(\mu, \Phi \otimes \Sigma), \qquad \Phi \in \mathscr{S}(n), \qquad \Sigma \in \mathscr{S}(p) \tag{1.3}$$

means that the $np \times 1$ vector $(u_1, \ldots, u_n)'$ is normally distributed with mean vector $(\mu_1, \ldots, \mu_n)'$ and covariance matrix $\Phi \otimes \Sigma$, where μ_i is the ith row of the $n \times p$ matrix μ, and $\Phi \otimes \Sigma$ is the Kronecker product of Φ and Σ.

Hence for $\Phi = (\phi_{ij})$, the covariance matrix of u_i and u_j in (1.3) is given by $\phi_{ij}\Sigma$: $p \times p$; thus, if $\Phi = I$, the n rows $u_1, \ldots, u_n$ of U are independent with a common covariance matrix Σ. Of course, $p = 1$ corresponds to the univariate case. The following elementary result on the normal distribution is left as an exercise (Exercise 1.2).

Lemma 1.1. *When $U: n \times p \sim N(\mu, \Phi \otimes \Sigma)$, then for any $A: m \times n$ and $B: q \times p$, $AUB' \sim N(A\mu B', A\Phi A' \otimes B\Sigma B')$.*

In most multivariate statistical analyses under normality, a data matrix $X: n \times p$ representing observations on n individuals or objects about p

characteristics is assumed to be distributed as

(1.4) $X \sim N(M, \Phi \otimes \Sigma), \qquad M \in R^{n \times p}, \qquad \Phi \in \mathscr{S}(n), \qquad \Sigma \in \mathscr{S}(p).$

Here, depending on the problem concerned, the mean matrix M has a specified structure. For example, $M = A\beta$ in the MANOVA model, which includes the case $M = \mathbf{1}\mu'$ with $\mathbf{1} = (1, \ldots, 1)' \in R^n$ and $\mu \in R^p$, and $M = A\beta C$ in the GMANOVA model, where β is an unknown matrix, and A and C are known matrices (see Chapters 3 and 5 for the models and their canonical forms). The dispersion matrix $\Phi \otimes \Sigma$ of X represents the assumption that the two rows x_i' and x_j' of the n rows of X have the covariance matrix

$$\mathrm{Cov}(x_i, x_j) = \phi_{ij}\Sigma \qquad (i, j = 1, \ldots, n),$$

where $\Phi = (\phi_{ij})$. In situations when $\Phi = I_n$, the n rows are independently distributed with a common covariance matrix Σ, and when $\Phi = I_n$ and $M = \mathbf{1}\mu'$, the different rows of X can be thought of as representing a homogeneous p-variate sample in general, and they are independently and identically distributed as $N_p(\mu, \Sigma)$.

1.2. Elliptically Symmetric Distributions

In this section, we first generalize the concept of spherical symmetry given in Definition 1.1 into that of elliptical symmetry; we then investigate some properties of the elliptically symmetric distribution and consider some implications of the assumption of the elliptical symmetry.

One of the most important properties of normal distribution is that the normality is preserved under linear transformations: for any $m \times p$ matrix A and $a \in R^m$,

(1.5) $x\colon p \times 1 \sim N(\mu, \Sigma) \Rightarrow Ax + a \sim N(A\mu + a, A\Sigma A'),$

where $\Sigma \in \mathscr{S}(p)$. This is the basic structure of normality, which provides for many testing problems the useful invariant structure (see Section 2.3 of Chapter 2). The property (1.5) does not hold for a nonnormal distribution with x_i's being independent where $x = (x_i)$. On the other hand, when we generalize the concept of spherical symmetry to that of elliptical symmetry, the property (1.5) holds with $A \in \mathrm{GL}(p)$ where $\mathrm{GL}(p)$ denotes the set of $p \times p$ nonsingular matrices. The definition of the elliptical symmetry of a distribution is given below.

Definition 1.3. Let x be a $p \times 1$ random vector. Then $\mathscr{L}(x)$ is said to be elliptically symmetric with location $\mu \in R^p$ and scale matrix $\Sigma \in \mathscr{S}(p)$ if

$\mathscr{L}(\Sigma^{-1/2}(x - \mu))$ is spherically symmetric in the sense of Definition 1.1, where $\mathscr{L}(x)$ denotes the distribution of x.

By $\mathscr{P}_E(\mu, \Sigma)$, we denote the class of elliptically symmetric distributions on R^p with location vector $\mu \in R^p$ and scale matrix $\Sigma \in \mathscr{S}(p)$. Of course, $\mathscr{P}_E(0, I_p)$ is the class of spherical symmetric distributions on R^p.

The following lemma corresponds to (1.5) with $A \in \mathrm{GL}(p)$ and normality replaced by an elliptically symmetric distribution. The proof is left as an exercise (Exercise 1.3).

Lemma 1.2. *If $\mathscr{L}(x) = P_{(\mu,\Sigma)}$ for some $P_{(\mu,\Sigma)} \in \mathscr{P}_E(\mu, \Sigma)$, then $\mathscr{L}(Ax + a) = P_{(A\mu + a,\, A\Sigma A')} \in \mathscr{P}_E(A\mu + a,\, A\Sigma A')$, where $A \in \mathrm{GL}(p)$ and $a \in R^p$.*

An example of an elliptically symmetric distribution is a distribution with the pdf (probability density function)

$$f(x \mid \mu, \Sigma) = |\Sigma|^{-1/2} q((x - \mu)'\Sigma^{-1}(x - \mu)) \tag{1.6}$$

where q is a function on $[0, \infty)$ satisfying $\int_{R^p} q(u'u)\,du = 1$. In fact, it is easy to see that whenever $\mathscr{L}(x) \in \mathscr{P}_E(\mu, \Sigma)$ has a pdf, the pdf is of the form (1.6) for some q (Exercise 1.4). Clearly the pdf of $N(\mu, \Sigma)$ is of the form (1.6) with $q(z) = (2\pi)^{-p/2}\exp(-z/2)$. Furthermore, if a pdf is of the form (1.6), so is its scale mixture

$$g(x \mid \mu, \Sigma) = \int_0^\infty a^{-p/2}|\Sigma|^{-1/2} q((x - \mu)'\Sigma^{-1}(x - \mu)/a)\,dG(a), \tag{1.7}$$

where $G(a)$ is a distribution on $(0, \infty)$. When the integrand q of (1.7) is a normal density, it is often called a normal mixture. A well-known example of a normal mixture is the multivariate t-distribution with degrees of freedom (d.f.) m whose pdf is given by

$$g(x \mid \mu, \Sigma) = c(m, p)|\Sigma|^{-1/2}\left[1 + \frac{1}{m}(x - \mu)'\Sigma^{-1}(x - \mu)\right]^{-(m+p)/2} \tag{1.8}$$

where $c(m, p) = \Gamma((m + p)/2)/(\pi m)^{p/2}\Gamma(m/2)$ and $\Gamma(a)$ denotes the gamma function. In fact, (1.8) follows from (1.7) with $dG(a) = [(m/2)^{m/2}a^{-m/2-1} \times \exp(-m/2a)/\Gamma(m/2)]\,da$, which is the pdf of the inverse gamma distribution (i.e., a^{-1} is gamma–distributed) (Exercise 1.5). In particular, the multivariate t-distribution with d.f. $m = 1$ is called multivariate Cauchy distribution for which no moments exist. The following lemma gives a condition for an elliptically symmetric distribution to be a normal mixture.

Lemma 1.3. (*Feller*, 1966, *p.* 439 *and* 441 *and Berger*, 1976.)

(1) *A necessary and sufficient condition for a* pdf *of the form* (1.6) *to be a normal mixture for some* $G(a)$ *is that* q *satisfies*

(1.9) $$(-1)^k[(d^k/ds^k)q(s)] \geq 0 \qquad (k = 1, 2, \ldots).$$

(2) *Suppose* q *satisfies* (1.9) *on* $(0, \infty)$ *and* $\psi(s) \geq 0$ *such that* $\psi'(s)$ *satisfies* (1.9). *Then* $q(\psi(s))$ *satisfies* (1.9).

Since this result is not used further in this book, the proof is left as an exercise. When q satisfies (1.9), q is called completely monotonic.

By using this lemma, a pdf such as

(1.10) $$f(x \mid \mu, \Sigma) = |\Sigma|^{-1/2} \exp\{-[(x-\mu)'\Sigma^{-1}(x-\mu)]^{1/2}\}/c(p)\Gamma(p)$$

can be shown to be a normal mixture, where $c(p)$ is the volume of the unit sphere $\{z \in R^p \mid \|z\| = 1\}$. Here $\|z\|^2 = z'z$.

We have seen above that the class $\mathscr{P}_E(\mu, \Sigma)$ of elliptically symmetric distributions contains a heavy-tail distribution such as the multivariate Cauchy distribution with no moments. Concerning the moments of a pdf of the form (1.6), the following results are easy to prove.

Lemma 1.4. (*Exercise* 1.7)

(1) *For a function of the form* (1.6) *to be a* pdf, $r^{p-1}q(r^2)$ *must be integrable on* $[0, \infty)$.

(2) *For a* pdf *of the form* (1.6) *to have the* mth *moments* $E(x_i^m)$ $(i = 1, \ldots, p)$, $r^{m+p-1}q(r^2)$ *must be integrable on* $[0, \infty)$.

(3) *If* $\mathscr{L}(x) \in \mathscr{P}_E(\mu, \Sigma)$ *and* $E|x_i| < \infty \quad (i = 1, \ldots, p)$, *then* $E(x) = \mu$.

(4) *If* $\mathscr{L}(x) \in \mathscr{P}_E(\mu, \Sigma)$ *and* $E\|x\|^2 < \infty$, *then* $\operatorname{Var}(x) = \alpha\Sigma$ *for some* $\alpha > 0$.

The following result, which is a slightly generalized version of Kelker (1970), gives another reason for the adoption of the assumption of elliptical symmetry in our discussion of robustness.

Lemma 1.5. *Let* $\mathscr{L}(x) = P_0$ *where* $x \in R^p$.

(1) $P_0 \in \mathscr{P}_E(\mu, \Sigma)$ *if and only if the characteristic function of* x *is of the form* $\exp(it'\mu)\psi(t'\Sigma t)$ *for some function* ψ *on* $[0, \infty)$ *where* $t \in R^p$.

(2) *Assume* $P_0 \in \mathscr{P}_E(\mu, \Sigma)$. *Let* $x = (x_1', x_2')'$ *with* x_i: $p_i \times 1$, $\mu = (\mu_1', \mu_2')'$ *with* μ_i: $p_i \times 1$, *and* $\Sigma = (\Sigma_{ij})$ *with* Σ_{ij}: $p_i \times p_j$ $(i, j = 1, 2)$. *Then the marginal distribu-*

tion of x_2 *belongs to* $\mathscr{P}_{\mathrm{E}}(\mu_2, \Sigma_{22})$, *which is functionally independent of* μ_1, Σ_{11} *and* Σ_{12}.

(3) *Assume* $P_0 \in \mathscr{P}_{\mathrm{E}}(\mu, \Sigma)$ *with* $\Sigma \in \mathscr{S}(p)$. *The conditional distribution of* x_1 *given* x_2 *belongs to* $\mathscr{P}_{\mathrm{E}}(\mu_1 + \Sigma_{12}\Sigma_{22}^{-1}(x_2 - \mu_2), \Sigma_{11.2})$ *where* $\Sigma_{11.2} = \Sigma_{11} - \Sigma_{12}\Sigma_{22}^{-1}\Sigma_{21}$.

Proof. (1) is left as an exercise. To prove (2), let $z_i = x_i - \mu_i\,(i = 1, 2)$, and let $c(t_1, t_2)$ be the characteristic function of (z_1, z_2). Then by (1)

$$c(t_1, t_2) = \psi\left((t_1', t_2')\Sigma\begin{pmatrix} t_1 \\ t_2 \end{pmatrix}\right) \tag{1.11}$$

for some ψ, since $\mathscr{L}((z_1', z_2')') \in \mathscr{P}_{\mathrm{E}}(0, \Sigma)$. Hence the characteristic function of z_2 is $c(0, t_2) = \psi(t_2'\Sigma_{22}t_2)$, implying the desired result by (1). To prove (3), let $y = z_1 - \Sigma_{12}\Sigma_{22}^{-1}z_2$, and consider the characteristic function of $\mathscr{L}(y, z_2)$:

$$\phi(u, v) = E[\exp(iu'y + iv'z_2)] = E[\exp(iu'z_1 + i(v - \beta u)'z_2)], \tag{1.12}$$

where $\beta = \Sigma_{22}^{-1}\Sigma_{21}$. Hence from (1.11)

$$\phi(u, v) = c(u, v - \beta u) = \psi(u'\Sigma_{11.2}u + v'\Sigma_{22}v).$$

This implies $\mathscr{L}((y', z_2')') \in \mathscr{P}_{\mathrm{E}}(0, \Omega)$ where $\Omega = (\Omega_{ij})$ with $\Omega_{11} = \Sigma_{11.2}$, $\Omega_{12} = 0$ and $\Omega_{22} = \Sigma_{22}$, which in turn implies $\mathscr{L}((w_1', w_2')') \in \mathscr{P}_{\mathrm{E}}(0, I_p)$ where $w_1 = \Sigma_{11.2}^{-1/2}y$ and $w_2 = \Sigma_{22}^{-1/2}z_2$. Therefore it is sufficient to show that the conditional distribution $\mathscr{L}(w_1 \mid w_2)$ of w_1 given w_2 belongs to $\mathscr{P}_{\mathrm{E}}(0, I_{p_1})$. But by the definition of conditional distribution and by the spherical symmetry of (w_1, w_2), for any Borel sets $A \subset R^{p_1}$ and $B \subset R^{p_2}$ and for any $\Gamma \in \mathscr{O}(p_1)$,

$$\begin{aligned}\int_{R^{p_1} \times B} Q(\Gamma A \mid w_2)P_1(dw_1, dw_2) &= \int_{\Gamma A \times B} P_1(dw_1, dw_2) = \int_{A \times B} P_1(dw_1, dw_2) \\ &= \int_{R^{p_1} \times B} Q(A \mid w_2)P_1(dw_1, dw_2)\end{aligned}$$

where $Q(\cdot \mid w_2)$ is the conditional distribution of w_1 given w_2 and P_1 is the distribution of (w_1, w_2) (which is in fact P_0 with $\mu = 0$ and $\Sigma = I$). This implies $Q(\Gamma A \mid w_2) = Q(A \mid w_2)$ (a.e. w_2) for any $\Gamma \in \mathscr{O}(p_1)$ and Borel set $A \subset R^{p_1}$, completing the proof.

Note that (3) can be proved immediately if the existence of the pdf is assumed (Exercise 1.9).

It follows from (3) that when $\mathrm{Cov}(x) = \alpha\Sigma$ exists, the conditional expectation and variance of x_1 given x_2 are, respectively, given by

(1.13) $$E[x_1 \mid x_2] = \mu_1 + \Sigma_{12}\Sigma_{22}^{-1}(x_2 - \mu_2) \quad \text{and} \quad \mathrm{Cov}(x_1 \mid x_2) = \alpha\Sigma_{11.2}.$$

Nimmo–Smith (1979) has shown that (x_1, x_2) cannot have the property (1.13) unless (x_1, x_2) is elliptically symmetric. The results (2) and (3) in Lemma 1.5 provide the same linear structure of the model as that of the normal distribution and play an important role in our subsequent analysis from an invariance viewpoint.

The following lemma is sometimes used in later chapters.

Lemma 1.6. *Let* $\mathscr{L}(x) = P \in \mathscr{P}_{\mathrm{E}}(0, I_p)$ *with* $P(x = 0) = 0$ *and* $\mathscr{L}(w) = N(0, I_p)$.

(1) $\mathscr{L}(x/\|x\|) = \mathscr{L}(w/\|w\|)$ *whatever* $\mathscr{L}(x) \in \mathscr{P}_{\mathrm{E}}(0, I_p)$ *may be.*

(2) *Let* $z = x/\|x\|$ *with* $z = (z_i)$. *Then* $\mathscr{L}((z_1^2, \ldots, z_p^2))$ *is the Dirichlet distribution* $D(\frac{1}{2}, \ldots, \frac{1}{2}, \frac{1}{2})$ *with* pdf

$$f(v_1, \ldots, v_p) = \Gamma(p/2)\left[\prod_{i=1}^{p-1} v_i^{1/2-1}\right](1 - v_1 - \cdots - v_{p-1})^{1/2-1} \Big/ \prod_{i=1}^{p} \Gamma(1/2)$$

where $0 \leqq v_i \leqq 1$ *and* $\Sigma_{i=1}^{p} v_i = 1$.

(3) $\mathscr{L}(z_1^2 + \cdots + z_{p_1}^2)$ *is beta distribution* $Be(p_1/2, (p - p_1)/2)$.

Proof. (1) Let U be a $p \times p$ random matrix such that $\mathscr{L}(U) = N(0, I_p \otimes I_p)$ and U and x are independent. Then $P(U \in \mathrm{GL}(p)) = 1$; hence, assume $U \in \mathrm{GL}(p)$. Let $\Psi = \Psi(U)$ be the orthogonal matrix obtained through the Gram–Schmidt orthogonalization such that $U = \Psi T$ with $T \in \mathrm{GU}(p)$, where $\mathrm{GU}(p)$ is the set of $p \times p$ upper triangular matrices with positive diagonal elements. Then the correspondence $(\Psi, T) \in \mathscr{O}(p) \times \mathrm{GU}(p) \to U \in \mathrm{GL}(p)$ is bijective (one-to-one, onto) and bimeasurable. Furthermore, $\Psi(\Gamma U) = \Gamma\Psi(U)$ for all $\Gamma \in \mathscr{O}(p)$ (Exercise 1.10). Since $\mathscr{L}(\Gamma U) = \mathscr{L}(U)$ for any $\Gamma \in \mathscr{O}(p)$, this implies

(1.14) $$\mathscr{L}(\Gamma\Psi(U)) = \mathscr{L}(\Psi(U)) \quad \text{for any} \quad \Gamma \in \mathscr{O}(p).$$

Now from $\mathscr{L}(\Phi x) = \mathscr{L}(x)$ for any $\Phi \in \mathscr{O}(p)$, with $z = x/\|x\|$

(1.15) $$\mathscr{L}(\Phi z) = \mathscr{L}(\Phi x/\|\Phi x\|) = \mathscr{L}(z) \quad \text{for any} \quad \Phi \in \mathscr{O}(p).$$

Therefore, by using the independence of U and x, we obtain

$$(1.16)\quad \begin{aligned} E[\exp(it'z)] &= E\{E[\exp(it'\Psi(U)'z)\,|\,U]\} = E[\exp(it'\Psi(U)'z)] \\ &= E\{E[\exp(it'\Psi(U)'z)\,|\,z]\} = E\{E[\exp(it'\Psi(U)'\Gamma'z)\,|\,z]\} \\ &= E[\exp(it'\Psi_1(U))], \end{aligned}$$

where $\Gamma \in \mathcal{O}(p)$ such that $\Gamma'z = (1, 0, \ldots, 0)'$, and $\Psi_1(U)'$ is the first row of $\Psi(U)$. In the first equality of (1.16), (1.15) is used, whereas in the fourth equality, (1.14) is used. Consequently, from the uniqueness of the characteristic function, $\mathscr{L}(x/\|x\|) = \mathscr{L}(\Psi_1(U))$, whatever $\mathscr{L}(x) \in \mathscr{P}_{\mathrm{E}}(0, I_p)$ may be, provided $P(x = 0) = 0$. Therefore since $\mathscr{L}(w) = N(0, I_p) \in \mathscr{P}_{\mathrm{E}}(0, I_p)$, $\mathscr{L}(x/\|x\|) = \mathscr{L}(w/\|w\|)$, completing the proof. Thus, (2) and (3) follow by assuming $\mathscr{L}(x) = N(0, I_p)$ by (1) (Exercise 1.11).

The result (1) in Lemma 1.6 is often stated as $z = x/\|x\|$ being uniformly distributed on the unit sphere $\mathscr{Z} = \{z \in R^p \,|\, \|z\| = 1\}$. This is because under the distribution P^z of z

$$P^z(\Gamma A) = P(z \in \Gamma A) = P(z \in A) = P^z(A) \qquad \text{for any} \quad \Gamma \in \mathcal{O}(p),$$

where A is a Borel set of $\mathscr{Z}$. That is, the volume of a surface area A is invariant under an orthogonal transformation (rotation or reflection) under the surface measure P^z. Lemma 1.6 (1) shows that such an invariant surface measure on $\mathscr{Z}$ is unique. This result is generalized in the next section.

There are many references on elliptically symmetric distributions; some basic articles are by Schoenberg (1938), Kelker (1970), Eaton (1981).

1.3. Left-Orthogonally Invariant Distributions

To fix the notation, for an $n \times p$ random matrix X, let

(1.17)

$$\mathscr{L}(X) \in \mathscr{P}_{\mathrm{E}}(\mu, \Phi \otimes \Sigma) \qquad \text{with} \quad \mu\colon n \times p, \qquad \Phi \in \mathscr{S}(n) \quad \text{and} \quad \Sigma \in \mathscr{S}(p)$$

mean $\mathscr{L}(x) \in \mathscr{P}_{\mathrm{E}}(\tilde{\mu}, \Phi \otimes \Sigma)$, where $x = (x_1, \ldots, x_n)'$ with x_i the ith row of X, $\tilde{\mu} = (\mu_1, \ldots, \mu_n)'$ with μ_i the ith row of μ. As before, $\mathscr{L}(\cdot)$ denotes the distribution of $(\cdot)$ and $\mathscr{S}(n)$ denotes the set of $n \times n$ positive definite matrices. For example, when $X\colon n \times p \sim N(\mu, \Phi \otimes \Sigma)$, as in (1.4), $\mathscr{L}(X) \in \mathscr{P}_{\mathrm{E}}(\mu, \Phi \otimes \Sigma)$.

The pdf of X in (1.17), when it exists, is of the form

$$f(X \mid \mu, \Sigma) = |\Phi|^{-p/2}|\Sigma|^{-n/2}q(\operatorname{tr}\Phi^{-1}(X-\mu)\Sigma^{-1}(X-\mu)') \tag{1.18}$$

for some q on $[0, \infty)$. In the case of $\mu = 0$ and $\Phi = I$, this density satisfies $f(\Gamma X \mid 0, \Sigma) = f(X \mid 0, \Sigma)$ for any $\Gamma \in \mathcal{O}(n)$ where $\mathcal{O}(n)$ is the set of $n \times n$ orthogonal matrices. Since the Jacobian of the transformation $X \to \Gamma X$ is 1, this implies that the distribution or a pdf of the form (1.18) with $\mu = 0$ and $\Phi = I$ is invariant under orthogonal transformations. This observation leads to the following definition.

Definition 1.4. An $n \times p$ random matrix X or its distribution is said to be left-orthogonally invariant about $\mu \in R^{n \times p}$ if $\mathcal{L}(\Gamma(X-\mu)) = \mathcal{L}(X-\mu)$ for all $\Gamma \in \mathcal{O}(n)$. The class of left-orthogonally invariant distributions about μ is denoted by $\mathcal{P}_{\mathrm{L}}(\mu)$.

When $p = 1$, the left-orthogonal invariance with $\mu = 0$ is reduced to the spherical symmetry in Definition 1.1. However when $p > 1$,

$$\mathcal{P}_{\mathrm{E}}(\mu, I_n \otimes \Sigma) \subseteq \mathcal{P}_{\mathrm{L}}(\mu) \qquad \text{for any} \quad \mu \in R^{n \times p} \quad \text{and} \quad \Sigma \in \mathcal{S}(p), \tag{1.19}$$

and the equality in (1.19) holds only if $p = 1$. Note that $\mathcal{P}_{\mathrm{E}}(\mu, aI_n) = \mathcal{P}_{\mathrm{E}}(\mu, I_n)$ for $a > 0$. Therefore for a matrix-variate random variable, $\mathcal{P}_{\mathrm{L}}(\mu)$ is broader than $\mathcal{P}_{\mathrm{E}}(\mu, I_n \otimes \Sigma)$ and hence the robustness under $\mathcal{P}_{\mathrm{L}}(\mu)$ is stronger than the one under $\mathcal{P}_{\mathrm{E}}(\mu, I_n \otimes \Sigma)$ for any $\Sigma \in \mathcal{S}(p)$.

An example of a distribution in $\mathcal{P}_{\mathrm{L}}(\mu)$ that is not elliptically symmetric is the matrix-variate t-distribution with pdf

$$h(X \mid \mu, \Sigma) = c|\Sigma|^{-n/2}|I + \Sigma^{-1}(X-\mu)'(X-\mu)|^{-(n+m)/2} \quad (m \geq p). \tag{1.20}$$

When $m = p$, the distribution with this pdf is sometimes called matrix-variate Cauchy distribution, for which no moments exist. Fraser and Ng (1980) treated a multivariate regression model with this distribution for the error term. In general, when $\mathcal{L}(X) \in \mathcal{P}_{\mathrm{L}}(\mu)$ has a density, it is of the form

$$f(X \mid \mu) = r((X-\mu)'(X-\mu)), \tag{1.21}$$

where $r(\cdot)$ is a nonnegative function on the set of $p \times p$ nonnegative definite matrices, say $\bar{\mathcal{S}}(p)$, such that $\int_{R^{n \times p}} r(X'X)\,dX = 1$. A mixture of this density

$$g(X \mid \mu) = \int_{\mathcal{S}(p)} r(A^{-1/2}(X-\mu)'(X-\mu)A^{-1/2})Q(dA) \tag{1.22}$$

is also of the form (1.21), where $Q(dA)$ is a probability measure on $\mathscr{S}(p)$. When $Q(dA)$ is the inverted Wishart measure (i.e., A^{-1} is Wishart-distributed), (1.22) with $r(C) \propto \exp(-\frac{1}{2}\hbar\, C)$ gives (1.20) with $\Sigma = I$.

Some properties of a distribution in $\mathscr{P}_{\mathrm{L}}(\mu)$ are described below.

Lemma 1.7. *Let $\mathscr{L}(X) \in \mathscr{P}_{\mathrm{L}}(\mu)$, where $X = (x_{ij})$: $n \times p$ and $\mu = (\mu_{ij})$: $n \times p$.*

(1) *For* $X = \begin{pmatrix} X_1 \\ X_2 \end{pmatrix}$ *with* X_i: $n_i \times p$, $\mathscr{L}(X_i) \in \mathscr{P}_{\mathrm{L}}(\mu_i)$, *where* $\mu = \begin{pmatrix} \mu_1 \\ \mu_2 \end{pmatrix}$ *with* μ_i: $n_i \times p$ $(i = 1, 2)$.

(2) *If* $E(|x_{ij}|) < \infty$ $(i = 1, \ldots, n,\ j = 1, \ldots, p)$, *then* $E(X) = \mu$.

(3) *If* $E[\operatorname{tr} X'X] < \infty$, *then* $\operatorname{Cov}(X) = I_n \otimes \Sigma$ *for some nonnegative definite matrix* Σ.

Proof. (1) and (2), respectively, are similar to (2) of Lemma 1.5 and (3) of Lemma 1.4, and are left as exercises (Exercise 1.12). To prove (3), let $Z = X - \mu$ and let z_i be the ith row of Z. Then the characteristic function of Z is of the form

$$c(T) = E[\exp(i \operatorname{tr} T'Z)] = \psi(T'T) \qquad \text{for some} \quad \psi, \tag{1.23}$$

where $T = (t_{ij})$: $n \times p$ because $\mathscr{L}(Z) \in \mathscr{P}_{\mathrm{L}}(0)$. Hence the characteristic function of z_i is of the form $\psi(t_i't_i)$ for all $i = 1, \ldots, n$, where t_i is the ith row of T. Hence all the covariance matrices of the z_i's are the same, say Σ, since the existence is assumed and $\Sigma = (-1)(\partial^2\psi(t_i't_i)/\partial t_{ij}\partial t_{ik}) \,|\, t_i = 0$. Also the characteristic function of a pair (z_i, z_j) $(i \neq j)$ is of the form $d(t_i, t_j) \equiv \psi(t_i't_i + t_j't_j)$ for all $i \neq j$ $(i, j = 1, \ldots, n)$. Here because $d(-t_i, t_j) = d(t_i, t_j)$, $E(-z_i'z_j) = \operatorname{Cov}(-z_i, z_j) = \operatorname{Cov}(z_i, z_j) = E(z_i'z_j)$, which implies $\operatorname{Cov}(z_i, z_j) = 0$. Thus $\operatorname{Cov}(X) = \operatorname{Cov}(Z) = I_n \otimes \Sigma$. This completes the proof.

Some other properties of left-$\mathscr{O}(n)$-invariant distributions are found in Dempster (1969), Dawid (1977), Jensen and Good (1981), and Eaton (1983).

Next we shall establish an important result corresponding to Lemma 1.6 (1). Define

$$\mathscr{Z} = \{Z\colon n \times p \,|\, Z'Z = I_p\} (\subset R^{n \times p}), \tag{1.24}$$

which is sometimes referred to as the Stiefel manifold in $R^{n \times p}$. In (1.24), $n \geq p$ is implied. Let $\mathscr{X} = \{X\colon n \times p \,|\, \operatorname{rank}(X) = p\}$, and let $\mathrm{GU}(p)$ denote the set of $p \times p$ upper triangular matrices with positive diagonal elements. Then, as is well known, the map

$$f_1: (\Gamma, U) \in \mathscr{Z} \times \mathrm{GU}(p) \to X = \Gamma U \in \mathscr{X} \tag{1.25}$$

is bijective (Exercise 1.13) and bimeasurable. The map

$$f_2: (\Psi, A) \in \mathscr{Z} \times \mathscr{S}(p) \to X = \Psi A \in \mathscr{X} \tag{1.26}$$

is also bijective and bimeasurable. Further, let $W: n \times n \sim N(0, I_n \otimes I_n)$ and decompose W into $W = VB$ with $V \in \mathcal{O}(n)$ and $B \in \mathrm{GU}(n)$ by the Gram–Schmidt orthogonalization procedure for the columns of W. Then $V = V(W)$ satisfies $V(QW) = QV(W)$ for all $Q \in \mathcal{O}(n)$.

Theorem 1.1. *Let X be an $n \times p$ random matrix in $\mathscr{X}$ such that $\mathscr{L}(X) \in \mathscr{P}_{\mathrm{L}}(0)$ and $P(X \in \mathscr{X}) = 1$, and let $W: n \times n \sim N(0, I_n \otimes I_n)$, independent of X. Define $V = V(W) \in \mathcal{O}(n)$ as above and decompose $V = [V_1, V_2]$ with $V_1: n \times p$.*

(1) *For $X = \Gamma U$ with $\Gamma \in \mathscr{Z}$ and $U \in \mathrm{GU}(p)$, $\mathscr{L}(\Gamma) = \mathscr{L}(V_1)$ for any $\mathscr{L}(X) \in \mathscr{P}_{\mathrm{L}}(0)$.*

(2) *For $X = \Psi A$ with $\Psi \in \mathscr{Z}$ and $A \in \mathscr{S}(p)$, $\mathscr{L}(\Psi) = \mathscr{L}(V_1)$ for any $\mathscr{L}(X) \in \mathscr{P}_{\mathrm{L}}(0)$.*

Proof. (1) The distribution of Γ is induced by that of X by the map f_1 in (1.25) as follows:

$$\begin{aligned} P^{\Gamma}(\mathscr{C}) &= P[(\Gamma, U) \in \mathscr{C} \times \mathrm{GU}(p)] = P[\Gamma U \in f_1(\mathscr{C} \times \mathrm{GU}(p))] \\ &= P[X \in f_1(\mathscr{C} \times \mathrm{GU}(p))] \end{aligned} \tag{1.27}$$

where $\mathscr{C}$ is a Borel set of $\mathscr{Z}$. By the definition of f_1, it follows that $f_1(Q\mathscr{C} \times \mathrm{GU}(p)) = Q f_1(\mathscr{C} \times \mathrm{GU}(p))$ for any $Q \in \mathcal{O}(n)$. Therefore, since $\mathscr{L}(QX) = \mathscr{L}(X)$ for any $Q \in \mathcal{O}(n)$, it follows that $P^{\Gamma}(Q\mathscr{C}) = P^{\Gamma}(\mathscr{C})$, i.e., $\mathscr{L}(Q\Gamma) = \mathscr{L}(\Gamma)$ for all $Q \in \mathcal{O}(n)$. Hence for given W, the characteristic function of Γ is

$$c(T) = E[\exp(i \operatorname{tr} T'\Gamma)] = E[\exp(i \operatorname{tr} T'V(W)'\Gamma) \mid W], \tag{1.28}$$

where $T: n \times p$. Taking the expectation with respect to W from both sides,

$$\begin{aligned} c(T) &= E[\exp(i \operatorname{tr} T'V(W)'\Gamma)] = E\{E[\exp(i \operatorname{tr} T'V(W)'\Gamma) \mid \Gamma]\} \\ &= E\{E[\exp(i \operatorname{tr} T'V(W)'Q'\Gamma) \mid \Gamma]\} = E[\exp(i \operatorname{tr} T'V_1(W)')], \end{aligned} \tag{1.29}$$

where $Q \in \mathcal{O}(n)$ satisfies $Q'\Gamma = \begin{pmatrix} I_p \\ 0 \end{pmatrix}$ for a given Γ, and $V_1(W) \in \mathscr{Z}$ is the submatrix in $V(W) = [V_1(W), V_2(W)]$. In the third equality of (1.29), the fact used is $\mathscr{L}(QV(W)) = \mathscr{L}(V(QW)) = \mathscr{L}(V(W))$ for all $Q \in \mathscr{O}(n)$. Therefore, by combining (1.28) and (1.29), the distribution of Γ is the same as that of V. The proof of (2) is similar and left as an exercise (Exercise 1.14).

Corollary 1.1. *Under the notation of Theorem* 1.1, $\mathscr{L}(Q\Gamma) = \mathscr{L}(\Gamma)$ *for all* $Q \in \mathscr{O}(n)$ *and* $\mathscr{L}(Q\Psi) = \mathscr{L}(\Psi)$ *for all* $Q \in \mathscr{O}(n)$.

Proof. The result follows from the proof of Theorem 1.1.

Some direct proofs of Theorem 1.1 are also found in Dempster (1969) and Eaton (1983).

Theorem 1.1 shows that (a) the distributions of $\Gamma = \Gamma(X)$ and $\Psi = \Psi(X)$ are the same; (b) the distribution of Γ remains the same whatever $\mathscr{L}(X)$ may be; (c) it is left $\mathscr{O}(n)$-invariant; and (d) the left-$\mathscr{O}(n)$-invariant distribution on $\mathscr{X}$ is the distribution on $\mathscr{X}$ induced by $V_1 = V_1(W)$ with $W \sim N(0, I_n \otimes I_n)$. In terms of Chapter 2, the above invariant distribution on $\mathscr{X}$ is described as the unique invariant measure on the homogeneous space $\mathscr{X}$ acted upon by the compact group $\mathscr{O}(n)$. The reader is referred to Nachbin (1965) and Eaton (1983) for a general proof of the existence and uniqueness of such an invariant measure on a homogeneous space.

Exercises

1.1. Prove the if-part of (1.1) in the following manner.

(a) Show that when $\mathscr{L}(x)$ is spherically symmetric, then the characteristic function $c(t)$ is a function of $\|t\|^2 = t_1^2 + \cdots + t_n^2$, say $\psi(\|t\|^2)$, where $t = (t_1, \ldots, t_n)' \in R^n$, and hence the characteristic function of x_i is given by $\psi(t_i^2)(i = 1, \ldots, n)$. [Hint: $c(t) = E(e^{it'x}) = E(e^{it'\Gamma x}) = c(\Gamma' t)$ for any $\Gamma \in \mathscr{O}(n)$.]

(b) Show that spherical symmetry with independence implies $c(t) = \exp(-\sigma^2\|t\|^2/2)$ for some $\sigma^2 > 0$. [Hint: $\psi(t_1^2 + \cdots + t_n^2) = \Pi_{i=1}^n \psi(t_i^2)$ for any $t_i \in R$. Hence with $\phi(s) = \log\psi(s)$, $\phi(s_1 + \cdots + s_n) = \Sigma_{i=1}^n \phi(s_i)$ for any $s_i > 0$, from which $\phi(s) = as$ follows for some $a > 0$.]

1.2. Prove Lemma 1.1.

1.3. Prove Lemma 1.2.

1.4. Prove that when $\mathscr{L}(x) \in \mathscr{P}_E(\mu, \Sigma)$ has a pdf (with respect to the Lebesgue measure), it is of the form (1.6). [Hint: First consider the case of $\mathscr{P}_E(0, I)$]

1.5. Derive (1.8) with $dG(a)$ given.

1.6 Show that the density in (1.10) is a normal mixture.

1.7 Prove Lemma 1.4.

1.8. Prove (1) in Lemma 1.5.

1.9 Prove (2) and (3) in Lemma 1.5 under the pdf (1.6).

1.10. In the proof of Lemma 1.6 (1), it was claimed that the map $(\Psi, T) \in \mathcal{O}(p) \times \mathrm{GU}(p) \to U \in \mathrm{GL}(p)$ is bijective and bimeasurable and that $\Gamma\Psi(U) = \Psi(\Gamma U)$ for any $\Gamma \in \mathcal{O}(p)$. Prove this claim.

1.11 Prove (2) and (3) in Lemma 1.6.

1.12 Prove (1) and (2) of Lemma 1.7.

1.13 Prove that the maps defined by (1.25) and (1.26) are bijective and bicontinuous (homeomorphisms).

1.14 Following the proof of Theorem 1.1 (1), prove Theorem 1.1 (2).

Chapter 2 Invariance Approach to Testing

2.1. Invariant Measures on Groups

In this chapter we review the invariance principle in the Neyman–Pearson–Wald testing theory and a powerful result for the derivation of the distribution of a maximal invariant due to Stein (1966), Wijsman (1967), and Andersson (1982). Here the term "invariance" is associated with the invariance structure of a testing problem that admits a transformation group. In multivariate hypotheses testing problems, such a group is usually a subset of a Euclidean space and so carries a topological structure in addition to the algebraic group structure. Consequently, we can make use of both structures in the analysis of a testing problem. In particular, the topological uniform structure compatible with the group structure enables us to utilize an invariant measure, known as the Haar measure, on the group. In this and the next section, some basic notions and results on this topic are presented, mostly without proofs. The reader is referred to Nachbin (1965), Berberian (1965), and Royden (1968) for mathematical details. The excellent book by Eaton (1983) is also very helpful for understanding the invariance principle in statistics. Some of the results of this chapter can also be found in a recent SIAM monograph by Eaton (1988).

We start with the definition of group.

Definition 2.1. A set $\mathscr{G}$ equipped with a binary operation $\circ: \mathscr{G} \times \mathscr{G} \to \mathscr{G}$ is said to be a group if

(i) for any g_1, g_2, and g_3 in $\mathscr{G}$, $(g_1 \circ g_2) \circ g_3 = g_1 \circ (g_2 \circ g_3)$,

(ii) there exists a unique element of $\mathscr{G}$, denoted by e, such that $e \circ g = g \circ e = g$ for all $g \in \mathscr{G}$,

(iii) for each $g \in \mathscr{G}$, there exists a unique element in $\mathscr{G}$, denoted by g^{-1}, such that $g \circ g^{-1} = g^{-1} \circ g = e$.

The element e is called the identity of $\mathscr{G}$, and g^{-1} is called the inverse of g.

For simplicity, we write $g_1 g_2$ for $g_1 \circ g_2$. A commonly used group is R^n with the binary operation addition $(+)$, where $0 \in R^n$ is the identity and the inverse of $x \in R^n$ is $-x$. The set $\mathrm{GL}(p)$ of all $p \times p$ nonsingular matrices, the set $\mathscr{O}(p)$ of all $p \times p$ orthogonal matrices, and the set $\mathrm{GU}(p)$ of all $p \times p$ upper triangular matrices with positive diagonal elements are all groups with matrix multiplication as the binary operations, where in each case the identity matrix I is the identity of the group, and the inverse matrix is the inverse element of the group. Usually $\mathrm{GL}(p)$ is called the general linear group. A group $\mathscr{G}$ is called *commutative* if $g_1 g_2 = g_2 g_1$ for any $g_1, g_2 \in \mathscr{G}$. R^n is commutative, while $\mathrm{GL}(p)$, $\mathscr{O}(p)$ and $\mathrm{GU}(p)$ are not commutative when $p \geqq 2$. $R_+ = \{x \in R \mid x > 0\}$ and $R_* = \{x \in R \mid x \neq 0\}$ are also commutative groups with multiplication as the binary operation $(\times)$, where 1 is the unit element for both R_+ and R_*. A group we often consider in statistics is the affine linear group $\mathrm{AL}(p) = \mathrm{GL}(p) \times R^p$, where the binary operation is defined by

$$(A_2, b_2) \circ (A_1, b_1) = (A_2 A_1,\ A_2 b_1 + b_2) \tag{2.1}$$

and the identity of $\mathrm{AL}(p)$ is $(I, 0)$.

Definition 2.2. A subset $\mathscr{H}$ of a group $(\mathscr{G}, \circ)$ is said to be a subgroup of $\mathscr{G}$ if $(\mathscr{H}, \circ)$ is a group.

Clearly $\mathscr{O}(p)$ and $\mathrm{GU}(p)$ are subgroups of $\mathrm{GL}(p)$, while R^p and $\mathrm{GL}(p)$ are subgroups of $\mathrm{AL}(p)$. Hence $\mathscr{O}(p)$, $\mathrm{GU}(p)$, $\mathrm{GL}(p)$, and R^p are all subgroups of $\mathrm{AL}(p)$. A feature of these groups is that they are regarded as subsets of Euclidean spaces. In fact, $\mathrm{GL}(p)$ is a subset of the set of $p \times p$ matrices, which is identified with $R^{p \times p}$. Also, $\mathrm{AL}(p)$ is a subset of $R^{p \times p} \times R^p = R^{p \times (p+1)}$. Hence these groups have the topologies induced from the Euclidean spaces, under which the binary operations are continuous. This leads to

Definition 2.3. A group $(\mathscr{G}, \circ)$ is said to be a topological group with topology $\mathscr{T}$ if under the topology $\mathscr{T}$ the map $(g_1, g_2) \in \mathscr{G} \times \mathscr{G} \to g_1 g_2^{-1} \in \mathscr{G}$ is continuous, where $\mathscr{G} \times \mathscr{G}$ is the topological product space.

As a topological space, $(\mathscr{G}, \mathscr{T})$ may be compact, locally compact, or sigma-compact. A topological space is called locally compact if each point has an open neighborhood with the compact closure, and it is called sigma-compact if it is expressed as a countable union of compact sets. In our examples, clearly the group $\mathscr{O}(p)$ is compact and the groups $\mathrm{GL}(p)$, $\mathrm{GU}(p)$ and $\mathrm{AL}(p)$ are locally compact (Exercise 2.2). To define a measure on a topological group $(\mathscr{G}, \mathscr{T})$, let $\mathscr{D}$ denote the smallest sigma-field generated by compact G_δ sets in $\mathscr{G}$ and let $\mathscr{B} = \mathscr{B}(\mathscr{T})$ be the Borel sigma-field generated by $\mathscr{T}$. Here a set is defined to be G_δ if it is expressed as a countable intersection of open sets. By the definitions, $\mathscr{D} \subset \mathscr{B}$.

Definition 2.4. A measure μ on a locally compact group $(\mathscr{G}, \mathscr{D})$ is said to be left-invariant if the following conditions (i) and (ii) hold:

(i) $\mu(gA) = \mu(A)$ for any $g \in \mathscr{G}$ and $A \in \mathscr{D}$.
(ii) $\mu(K) < \infty$ for any compact subset K.

Of course, a right invariant measure ν on $(\mathscr{G}, \mathscr{D})$ is similarly defined by replacing gA by Ag in (i). Such an invariant measure is often called a Haar measure.

Theorem 2.1. *There exists a left-invariant (right-invariant) measure μ (respectively, ν) on every locally compact group $(\mathscr{G}, \mathscr{D})$, and it is unique up to constant multiplications.*

Proof. See Nachbin (1965) and Berberian (1965), in which the result is shown in terms of integrals.

If $\mathscr{G}$ is sigma-compact, the invariant measure μ on $(\mathscr{G}, \mathscr{D})$ is regular so that it is uniquely extended to the Borel measure $\tilde{\mu}$ on $(\mathscr{G}, \mathscr{B})$, such that $\tilde{\mu} = \mu$ on $\mathscr{D}$ and $\tilde{\mu}$ satisfies certain conditions (see Royden, 1968, p. 307, 314). Therefore we may say there exists a unique invariant measure $\tilde{\mu}$ on $(\mathscr{G}, \mathscr{B})$, although the invariance holds only for the sets in $\mathscr{D}$, as in Definition 2.4. However, the groups we will treat in our applications are all separable locally compact metric spaces, in which $\mathscr{D} = \mathscr{B}$ holds. Consequently, the distinction between $\mathscr{D}$ and $\mathscr{B}$ vanishes. Furthermore, in our cases the invariant measures are directly constructed as will be seen below, and hence the existence in Theorem 2.1 is not used explicitly.

Now for a given left-invariant measure μ, define the integrals by

$$I(f) = \int_{\mathscr{G}} f(x)\mu(dx) \qquad \text{for} \quad f \in \mathscr{L}(\mathscr{D}), \tag{2.2}$$

where $\mathscr{L}(\mathscr{D})$ denotes the class of μ-integrable functions on $\mathscr{G}$. Also define for $a \in \mathscr{G}$ and $f \in \mathscr{L}(\mathscr{D})$,

$$(af)(x) = f(a^{-1}x) \quad \text{and} \quad (fa)(x) = f(xa^{-1}). \tag{2.3}$$

Then by the left-invariance of μ

$$I(af) = I(f) \qquad \text{for all} \quad a \in \mathscr{G} \tag{2.4}$$

(see Exercise 2.4), but $I(fa) = I(f)$ does not always hold because μ is not necessarily right-invariant.

Theorem 2.2. *Let μ be a left-invariant measure on a locally compact group $\mathscr{G}$. Then there exists a continuous homomorphism Δ from $\mathscr{G}$ into the group of positive reals $R_+ = \{x \in R \mid x > 0\}$, such that $\mu(Ag) = \Delta(g)\mu(A)$ for any $A \in \mathscr{D}$ and $g \in \mathscr{G}$.*

Proof. See Nachbin (1965, Section 5, Chapter 2) or Segal and Kunze (1968, p. 194).

It is noted that a map Δ from a group $(\mathscr{G}, \circ)$ into a group $(\mathscr{G}', *)$ is called a homomorphism if $\Delta(g_1 \circ g_2) = \Delta(g_1) * \Delta(g_2)$ so that Δ in Theorem 2.2 satisfies $\Delta(g_1 g_2) = \Delta(g_1)\Delta(g_2)$, $\Delta(e) = 1$, and $\Delta(g^{-1}) = (\Delta(g))^{-1}$. The homomorphism Δ in Theorem 2.2 is said to be the right modular function of the left-invariant measure μ. Since μ is unique up to constant multiplications, so is Δ. When $\Delta(g) \equiv 1$, $\mathscr{G}$ or μ is said to be *unimodular*. If μ is left- and right-invariant, then it is unimodular. By this theorem, it follows that $I(fa) = \Delta(a)I(f)$.

Lemma 2.1. (1) *$\mu(\mathscr{G}) < \infty$ if and only if $\mathscr{G}$ is compact.*
(2) *If $\mathscr{G}$ is compact, $\mathscr{G}$ is unimodular so that μ is left-and right-invariant.*

Proof. See Nachbin (1965) or Berberian (1965).

The result (1) implies that there exists no invariant probability measure on $\mathscr{G}$ unless $\mathscr{G}$ is compact.

Before we give some examples of invariant measures, we make one more definition.

Definition 2.5. A measure μ on a locally compact group $(\mathscr{G}, \mathscr{D})$ is said to be a relatively left-invariant measure with (left) multiplier χ if μ satisfies (i) $\mu(gA) = \chi(g)\mu(A)$ for any $A \in \mathscr{D}$ and $g \in \mathscr{G}$ and (ii) $\mu(K) < \infty$ for any compact set K.

Lemma 2.2. (1) *The multiplier χ is a continuous homomorphism from $\mathscr{G}$ into R_+.*
(2) *If μ is relatively left-invariant with multiplier χ, $\chi(x^{-1})\mu(dx)$ is left-invariant.*

Proof. See Nachbin (1965) for (1). For (2), define $J(f) = \int f(x)\chi(x^{-1})\mu(dx)$ for $f \in \mathscr{L}(\mathscr{D})$. Then

$$J(af) = \int f(a^{-1}x)\chi(x^{-1})\mu(dx) = \int f(a^{-1}x)\chi(a^{-1}ax^{-1})\mu(dx)$$

$$= \chi(a^{-1}) \int f(a^{-1}x)\chi((a^{-1}x)^{-1})\mu(dx)$$

$$= \chi(a^{-1})\chi(a) \int f(x)\chi(x^{-1})\mu(dx) = J(f).$$

This implies that J is a left-invariant integral on $\mathscr{L}(\mathscr{D})$, completing the proof.

Since the map $a: x \to ax$ defines a homeomorphism from $\mathscr{G}$ onto $\mathscr{G}$, the inverse of the multiplier χ in Definition 2.5 is interpreted as the "Jacobian" of the transformation. A relatively left-invariant measure is a measure that makes the "Jacobian" independent of x.

Now let us see some examples of invariant measures.

1. R^p: The Lebesgue measure dx is left- and right-invariant.
2. GL(p): First note that GL(p) $\subset R^{p \times p}$ and $R^{p \times p} -$ GL(p) has Lebesgue measure 0. Let $A(X) = AX$ for $A, X \in$ GL(p). Then the Jacobian of $X \to A(X)$ is $|A'A|^{-p/2}$. Hence by Lemma 2.2(2), $\mu(dX) = |X'X|^{-p/2}\, dX$ is left invariant. Because the Jacobian of $X \to XB$ is $|B'B|^{-p/2}$, it is right invariant too, though GL(p) is not commutative. Hence GL(p) is unimodular.
3. AL(p)$=$GL(p)$\times R^p$: Consider the transformation $(X, y) \to (A, b)(X, y) = (AX, Ay + b)$ and note that AL(p) $\subset R^{p \times p} \times R^p$ and $R^{p \times p} \times R^p -$ AL(p) has Lebesgue measure 0. Then, by computing the Jacobian, $\mu(dX, dy) = |X'X|^{-(p+1)/2}\, dX\, dy$ is shown to be left-invariant. Finding a right-invariant measure v is given as an exercise (Exercise 2.8). Note $v \neq \mu$.
4. GU(p): In a similar manner, $\mu(dX) = [\Pi x_{ii}^{p-i+1}]^{-1}\, dX$ is shown to be left-invariant, when $X = (x_{ij})$.
5. GT(p), the group of $p \times p$ lower triangular matrices with positive diagonal elements: $\mu(dX) = [\Pi_{i=1}^{p} x_{ii}^{i}]^{-1}\, dX$, is shown to be left invariant.
6. Define the group that is used in Chapter 5,

$$\text{(2.5)} \quad \mathscr{K} = \left\{A = \begin{pmatrix} A_{11} & A_{12} \\ 0 & A_{22} \end{pmatrix} : p \times p \,\middle|\, A_{ii} \in \mathrm{GL}(p_i) (i = 1, 2)\right\} \subset R^{p_1^2 + p_2^2 + p_1 p_2}.$$

Then $\mu(dA) = |A'_{11}A_{11}|^{-p/2}|A'_{22}A_{22}|^{-p_2/2}\,dA$ is shown to be left invariant by computing the Jacobian of $X \to AX$.

As has been seen in the proof of Theorem 1.1 of Chapter 1, the construction of the unique invariant probability measure on the group $\mathcal{O}(p)$ of $p \times p$ orthogonal matrices can be made through a normal distribution. That is, let $U \sim N(0, I_p \otimes I_p)$, and factor U into $U = \Gamma W$ with $\Gamma \in \mathcal{O}(p)$ and $W \in \mathrm{GU}(p)$ by the Gram–Schmidt orthogonalization procedure. Here $U \in \mathcal{X} = \{U: p \times p \,|\, \mathrm{rank}\,(U) = p\}$, since $P(U \in \mathcal{X}) = 1$. Then the decomposition is unique, and $\Gamma \equiv \Gamma(U)$ satisfies $Q\Gamma(U) = \Gamma(QU)$ for any $Q \in \mathcal{O}(p)$. Therefore the distribution of Γ satisfies

$$P^{\Gamma}(Q\mathscr{C}) = P(\Gamma(U) \in Q\mathscr{C}) = P(\Gamma(Q'U) \in \mathscr{C}) = P^{\Gamma}(\mathscr{C})$$

for all Borel sets $\mathscr{C} \subset \mathcal{O}(p)$ and $Q \in \mathcal{O}(p)$, showing that P^{Γ} is the invariant probability measure on $\mathcal{O}(p)$. Further, by its construction, the first column γ_1 of Γ is $u_1/\|u_1\|$ where u_1 is the first column of U. Consequently the distribution of $\gamma_1 \equiv (\gamma_{11}, \ldots, \gamma_{p1})'$ is described by that of $(u_{11}/\|u_1\|, \ldots, u_{p1}/\|u_1\|)'$ where $U = (u_{ij})$ and $\Gamma = (\gamma_{ij})$, and the expected values of the elements of a random orthogonal matrix are often computed by this relation. In particular, $(\gamma_{i1}^2, \ldots, \gamma_{ip}^2)$ has a Dirichlet distribution $D(\frac{1}{2}, \ldots, \frac{1}{2}, \frac{1}{2})$ (see Section 1.2, Chapter 1) and hence $E[(\gamma_{ij})^{2n}(\gamma_{ik})^{2n}]$ is easily evaluated. Some integral formulae on $\mathcal{O}(p)$ are listed below.

Lemma 2.3. (*Exercise* 2.10). *Let B be a $p \times p$ real matrix. Suppose $\Gamma B \Gamma' = B$ for all $\Gamma \in \mathcal{O}(p)$. Then $B = \alpha I$ for some $\alpha \in R$.*

Lemma 2.4. *Let ν be the invariant probability measure on $\mathcal{O}(p)$. Then for $\Gamma = (\gamma_{ij})$,*

$$\int_{\mathcal{O}(p)} \gamma_{ij}\gamma_{kl}\,\nu(d\Gamma) = \tfrac{1}{p}\delta_{ik}\delta_{jl}$$

where δ_{ij} is Kronecker's δ.

Proof. For any $p \times p$ matrix C, let $B = \int_{\mathcal{O}(p)} \Psi C \Psi' \nu(d\Psi)$. Then using the left-invariance of ν, it follows that $\Gamma B \Gamma' = B$ for all $\Gamma \in \mathcal{O}(p)$. Thus by Lemma 2.3, $B = \alpha I_p$. To determine α, take the trace of B; $\mathrm{tr}\,B = p\alpha = \int_{\mathcal{O}(p)} \mathrm{tr}\,\Psi C \Psi' \nu(d\Psi) = \mathrm{tr}\,C$. Hence $\alpha = \mathrm{tr}\,C/p$. Taking $c_{k1} = 0$ in $C = (c_{ij})$ for all $(k, 1)$ except for the (i, j) element c_{ij} and setting $c_{ij} = 1$, we obtain $\alpha = \frac{1}{p}\mathrm{tr}\,C = \frac{1}{p}\delta_{ij}$. Now comparing both sides of $\alpha I_p = \int_{\mathcal{O}(p)} \Psi C \Psi' \nu(d\Psi)$ with $\alpha = \frac{1}{p}\delta_{ij}$, we obtain the desired result.

Lemma 2.5. *Let A and B be $p \times p$ matrices. Then*

$$\int_{\mathcal{O}(p)} \operatorname{tr} A\Gamma B\Gamma' \nu(d\Gamma) = \tfrac{1}{p} \operatorname{tr} A \operatorname{tr} B$$

$$\int_{\mathcal{O}(p)} \operatorname{tr} A\Gamma \operatorname{tr} B\Gamma \nu(d\Gamma) = \tfrac{1}{p} \operatorname{tr} AB'.$$

The proof is left as an exercise (Exercise 2.11). See also James (1964).

2.2. Invariant Measures on Homogeneous Spaces

In this section, some notions of a transformation group and associated mathematical results are summarized. A more intensive argument on this topic is found in Eaton (1983) and Eaton (1988).

Definition 2.6. A group $(\mathcal{G}, \circ)$ is said to act on the left of a set $\mathcal{X}$ if there exists a map $T: \mathcal{G} \times \mathcal{X} \to \mathcal{X}$ such that (i) $T(g_2, T(g_1, x)) = T(g_2 \circ g_1, x)$ and (ii) $T(e, x) = x$ for any $g_1, g_2 \in \mathcal{G}$ and $x \in \mathcal{X}$.

Usually the value of $T(g, x)$ in the above definition is denoted by gx, and $T(g, x)$ is abbreviated as (g, x). In this notation, (i) and (ii) become (i) $(g_2, g_1 x) = (g_2 \circ g_1, x)$ and (ii) $(e, x) = x$, respectively.

Lemma 2.6. *For each $g \in \mathcal{G}$, define a map $\mathcal{X} \to \mathcal{X}$ by $T_g(x) \equiv T(g, x)$. Then T_g is bijective (one-to-one, onto).*

The proof is given as Exercise 2.13.

We shall often use the notation $(\mathcal{G}, \mathcal{X}, T)$ for the fact that a group $\mathcal{G}$ acts on the left of $\mathcal{X}$ with action T. A trivial example of $(\mathcal{G}, \mathcal{X}, T)$ is $(\mathcal{G}, \mathcal{G}, \circ)$ where $\circ$ is the binary operation on $\mathcal{G}$; $\circ\ (g, x) \to g \circ x$. The matrix groups $\mathrm{GL}(p)$, $\mathcal{O}(p)$, $\mathrm{GU}(p)$, etc., act on the left of R^n with $T(A, x) = Ax$. Also, $(\mathrm{AL}(p), R^n, T)$ is another example where

$$T((A, b), x) = Ax + b \quad \text{for} \quad (A, b) \in \mathrm{AL}(p), \quad x \in R^n. \tag{2.6}$$

Recall that $\mathrm{AL}(p) = \mathrm{GL}(p) \times R^p$.

Definition 2.7. Assume a particular $(\mathcal{G}, \mathcal{X}, T)$. Then $\mathcal{G}$ is said to act transitively on the left of $\mathcal{X}$ if for any x_1, $x_2 \in \mathcal{X}$, there exists $g_{12} \in \mathcal{G}$ such that

$T(g_{12}, x_1) = x_2$. Also $\mathscr{G}$ is said to act freely on the left of $\mathscr{X}$ if for any $x_1, x_2 \in \mathscr{G}$ there exists a unique element $g_{12} \in \mathscr{G}$ such that $T(g_{12}, x_1) = x_2$.

For a given $(\mathscr{G}, \mathscr{X}, T)$, fix $x \in \mathscr{X}$ and define the map

$$\pi_x(g) \equiv T(g, x)\colon \mathscr{G} \to \mathscr{X} \tag{2.7}$$

and the *orbit* of x,

$$\mathscr{G}x = \{gx \in \mathscr{X} \mid g \in \mathscr{G}\}. \tag{2.8}$$

When $\mathscr{G}$ acts transitively, it is easy to see that π_x in (2.7) is surjective (onto) and $\mathscr{G}x = \mathscr{X}$ (Exercise 2.15). But π_x is not always injective (one-to-one). In fact $\pi_x(g_1) = \pi_x(g_2)$ implies $g_1 x = g_2 x$ or $g_2^{-1} g_1 x = x$. Hence, unless $g_2^{-1} g_1 = e$, π_x is not injective. In other words, unless the *isotropy subgroup* defined by

$$\mathscr{H}_x = \{g \in \mathscr{G} \mid gx = x\} \tag{2.9}$$

is trivial, i.e., $\mathscr{H}_x = \{e\}$, π_x is not injective. In fact, $\mathscr{H}_x = \pi_x^{-1}(x)$. If $\mathscr{H}_x = \{e\}$ for one x, $\mathscr{H}_x = \{e\}$ for all x, since $\mathscr{G}$ is transitive. Hence $\mathscr{G}$ acts freely. When $\mathscr{G}$ acts freely, π_x is bijective for any x.

So far, we have not assumed topologies on $\mathscr{G}$ and $\mathscr{X}$. From now on, *we assume that $\mathscr{G}$ is a topological group and $\mathscr{X}$ is a Hausdorff topological space.*

Definition 2.8. For a given $(\mathscr{G}, \mathscr{X}, T)$, $\mathscr{G}$ is said to act topologically on the left of $\mathscr{X}$ if $T\colon \mathscr{G} \times \mathscr{X} \to \mathscr{X}$ is continuous. In such a case, we shall say that $(\mathscr{G}, \mathscr{X}, T)$ is topological. Also when $(\mathscr{G}, \mathscr{X}, T)$ is topological, $\mathscr{X}$ is said to be the homogeneous space of $\mathscr{G}$ if $\mathscr{G}$ acts transitively.

Lemma 2.7. *Suppose $(\mathscr{G}, \mathscr{X}, T)$ is topological.*

(1) *$T_g(x) = T(g, x)\colon \mathscr{X} \to \mathscr{X}$ is a homeomorphism for each $g \in \mathscr{G}$.*

(2) *$\pi_x(g) = T(g, x)\colon \mathscr{G} \to \mathscr{X}$ is continuous for each $x \in \mathscr{X}$.*

(3) *If $\mathscr{G}$ is transitive, π_x is open and surjective, i.e., π_x maps an open set onto an open set.*

The proof is left as an exercise (Exercise 2.17).

When $\mathscr{G}$ is free, then by (2) and (3), π_x is a homeomorphism. Therefore, assuming that $\mathscr{G}$ is locally compact, $\mathscr{X}$ is locally compact, and the sigma-field $\mathscr{D}$ on $\mathscr{G}$ generated by compact G_δ sets of $\mathscr{G}$ corresponds to the one $\tilde{\mathscr{D}}$ on $\mathscr{X}$ generated by those of $\mathscr{X}$. Hence, by letting μ be a left-invariant measure on $(\mathscr{G}, \mathscr{D})$ (see Theorem 2.1), the measure $\tilde{\mu}$ on $(\mathscr{X}, \tilde{\mathscr{D}})$ induced by μ,

$$\tilde{\mu}(B) = \mu(\pi_x^{-1}(B)), \qquad B \in \tilde{\mathscr{D}} \tag{2.10}$$

is an invariant measure on $(\mathscr{X}, \tilde{\mathscr{D}})$ under the action of $\mathscr{G}$. In fact, since there exists a unique $A \in \mathscr{D}$ such that $B = \pi_x(A)$, and since $gB = \pi_x(gA)$ for any $g \in \mathscr{G}$ by the definition of π_x, we obtain

$$(2.11) \qquad \tilde{\mu}(gB) = \mu(gA) = \mu(A) = \tilde{\mu}(B) \qquad \text{for any} \quad g \in \mathscr{G},$$

with $\tilde{\mu}(B) < \infty$ for a compact B. Further, by using the uniqueness of μ on $\mathscr{G}$, it is easily shown that the invariant measure $\tilde{\mu}$ on $(\mathscr{X}, \tilde{\mathscr{D}})$ is unique.

The above result followed from the existence of a homeomorphism $\psi: \mathscr{G} \to \mathscr{X}$ with $\psi(g_1 g) = g_1 \psi(g)$. However, when $\mathscr{G}$ is simply transitive, π_x is not injective as has been discussed above, and $\mathscr{G}$ cannot be homeomorphic to $\mathscr{X}$. But we know that $\pi_x^{-1}(x) = \mathscr{H}_x$ with $\mathscr{H}_x$ in (2.9) and $\mathscr{H}_x$ is a subgroup of $\mathscr{G}$. Hence we may question whether $\mathscr{G}/\mathscr{H}_x$ is homeomorphic to $\mathscr{X}$, where $\mathscr{G}/\mathscr{H}_x$ is the left quotient space with the induced topology each point of which is of the form $g\mathscr{H}_x$ with $g \in \mathscr{G}$. The following lemma answers the question positively.

Lemma 2.8. *For a topological* $(\mathscr{G}, \mathscr{X}, T)$, $\phi: g\mathscr{H}_x \in \mathscr{G}/\mathscr{H}_x \to gx \in \mathscr{X}$ *is a homeomorphism with* $\phi(g_1\mathscr{H}_x g_2 \mathscr{H}_x) = g_1 \phi(g_2 \mathscr{H}_x)$.

The proof is left as an exercise (see Nachbin, 1965, p. 133).

The following result due to Weil (see Nachbin (1965)) shows the existence and uniqueness of an invariant measure on a homogeneous space $\mathscr{X}$.

Theorem 2.3. *When* $(\mathscr{G}, \mathscr{X}, T)$ *is topological with* $\mathscr{G}$ *locally compact and* $\mathscr{X}$ *locally compact Hausdorff, in order that there exist an invariant measure* $\tilde{\mu}$ *on* $(\mathscr{X}, \mathscr{D})$ *under* $\mathscr{G}$ *that is unique except for a positive constant multiplication, it is necessary and sufficient that*

$$(2.12) \qquad \Delta_{\mathscr{H}}(h) = \Delta_{\mathscr{G}}(h) \qquad \text{for} \quad h \in \mathscr{H},$$

where $\Delta_{\mathscr{H}}$ *and* $\Delta_{\mathscr{G}}$ *are the right modular functions of* $\mathscr{H}$ *and* $\mathscr{G}$, *respectively.*

Proof. See Nachbin (1965, p. 138).

It is noted that $\mathscr{X}$ is Hausdorff if and only if $\mathscr{H}_x$ is closed.

This result is related to the result in Theorem 1.1 of Chapter 1, which includes the result of Lemma 1.5 of Chapter 1 as a special case. Let us consider it again in the present context.

Example 2.1. As in (1.24) of Chapter 1, define the Stiefel manifold

$$(2.13) \qquad \mathscr{Z} = \{Z: n \times p \,|\, Z'Z = I_p\}$$

and let $\mathcal{O}(n)$ act on $\mathcal{Z}$ by $T(\Gamma, Z) = \Gamma Z$, where $\Gamma \in \mathcal{O}(n)$ and $Z \in \mathcal{Z}$. Then, as is easily shown, $\mathcal{O}(n)$ acts transitively and $(\mathcal{O}(n), \mathcal{Z}, T)$ is topological with $\mathcal{O}(n)$ and $\mathcal{Z}$ compact. Since $\mathcal{Z}$ is Hausdorff,

$$\mathcal{H}_{Z_0} = \{\Gamma \in \mathcal{O}(n) \mid \Gamma Z_0 = Z_0\}, \tag{2.14}$$

where $Z_0 = \begin{pmatrix} I_p \\ 0 \end{pmatrix}$, is closed and thus compact. This implies $\Delta_{\mathcal{H}} \equiv 1$ by Lemma 2.1 (2), while $\Delta_{\mathcal{O}(n)} \equiv 1$. Therefore, by Theorem 2.3 there exists a unique (up to constant multiplications) invariant measure $\tilde{\mu}$ on $\mathcal{Z}$. This measure was constructed in Theorem 1.1 of Chapter 1 by using normal distribution. In particular, there exists a unique invariant probability measure on $\mathcal{Z}$. In the case of $p = 1$, $\mathcal{Z}$ is the sphere in R^n.

2.3. A Review of the Theory of Testing of Hypotheses

In this section, the basic framework of the Neyman–Pearson theory for the testing of statistical hypotheses is reviewed with special emphasis on invariance. The reader is referred to the classic textbook by Lehmann (1959) for details. Let

$$\mathcal{F}(\Theta) = \{f(x \mid \theta) \mid \theta \in \Theta\} \qquad (x \in \mathcal{X}) \tag{2.15}$$

be a class of probability densities on a subset $\mathcal{X}$ of a Euclidean space with respect to a sigma-finite measure μ, where the functional form of f is known, but f is parametrized by an unknown vector θ in a subset Θ of a Euclidean space. Here $\mathcal{F}(\Theta)$ is assumed to be identifiable (i.e., $f(x \mid \theta_1) = f(x \mid \theta_2)$, for almost all x implies $\theta_1 = \theta_2$). Then a testing problem is described by two disjoint subsets of Θ, say Θ_0 and Θ_1, as follows:

$$\text{H}: \theta \in \Theta_0 \qquad \text{versus} \qquad \text{K}: \theta \in \Theta_1 \tag{2.16}$$

where H is considered a maintained or null hypothesis to be tested against the alternative hypothesis K. A decision to make here is either to accept the null hypothesis H or to reject H relative to the alternative K on the basis of an observation x. A decision function, which we shall call a test or a test function below, is a measurable function $\phi(x)$ from $\mathcal{X}$ into $[0, 1]$, denoting the probability that H is rejected when x is observed. In particular, if $\phi(x) = 1$, for some x, H is rejected for that x. The average probability of a test for rejection under θ is defined by

$$\pi(\phi, \theta) = E_\theta[\phi(x)] = \int_{\mathcal{X}} \phi(x) f(x \mid \theta) \mu(dx), \tag{2.17}$$

and this function is called the power function of a test ϕ when $\theta \in \Theta_1$ or when K is true, meaning the average probability of the correct decision under a test ϕ when K is true. As is well known, in the Neyman–Pearson theory, we wish to find a test which maximizes $\pi(\phi, \theta)$ under $\theta \in \Theta_1$ in some sense in the class $\mathscr{D}$ of level α tests defined by

$$\mathscr{D} = \left\{\phi \mid \phi \text{ is a test}, \quad \sup_{\theta \in \Theta_0} \pi(\phi, \theta) \leq \alpha\right\} \qquad (0 < \alpha < 1). \tag{2.18}$$

Here the number α is called a significance level, and it corresponds to the maximum average probability of the incorrect decision under the test ϕ when H is true. The size of a test is defined to be the number $\sup_{\theta \in \Theta_0} \pi(\phi, \theta)$. In maximizing power, the following generalized Neyman–Pearson lemma is often used.

Lemma 2.9. (*Lehmann*, 1959, *p.* 83). *Let $f_1, \ldots, f_{m+1}$ be real-valued functions defined on $\mathscr{X}$ and integrable with respect to μ, and suppose that for given constants $c_1, \ldots, c_m$, there exists a test function ϕ satisfying*

$$\int \phi f_i \, d\mu = c_i \qquad (i = 1, \ldots, m). \tag{*}$$

Let $\mathscr{C}$ be the class of tests for which (*) holds. Then there exists a test ϕ in $\mathscr{C}$ that maximizes

$$\int \phi f_{m+1} \, d\mu. \tag{**}$$

If a test ϕ in $\mathscr{C}$ is of the form

$$\phi(x) = \begin{cases} 1 & \text{if} \quad f_{m+1}(x) > \sum_{i=1}^{m} k_i f_i(x) \quad \text{a.e.} \\ 0 & \text{if} \quad f_{m+1}(x) < \sum_{i=1}^{m} k_i f_i(x) \quad \text{a.e.} \end{cases} \tag{***}$$

for some constants $k_1, \ldots, k_m$, then it maximizes (**) in $\mathscr{C}$. If a test ϕ in *$\mathscr{C}$ satisfies (***) with $k_i \geq 0$ $(i = 1, \ldots, m)$, it maximizes (**) in the class of tests satisfying*

$$\int \phi_0 f_i \, d\mu \leq c_i \qquad (i = 1, \ldots, m).$$

If $m = 1$ with $f_1(x) = f(x \mid \theta_0)$ and $f_2(x) = f(x \mid \theta_1)$, where $\theta_0 \in \Theta_0$ and $\theta_1 \in \Theta_1$ are fixed, the above lemma is reduced to the Neyman–Pearson fundamental lemma.

However, except for certain specific cases, there exists no test that maximizes the power $\pi(\phi, \theta)$ uniformly in $\theta \in \Theta_1$ in the class of level α tests $\mathscr{D}$. An approach to a solution to this indeterminacy is to restrict the class $\mathscr{D}$ to a subclass of $\mathscr{D}$ by a certain criterion, and to try to find a test that maximizes the power as much as possible within this subclass or whose power function behaves well for the purpose of an analysis. Criteria to restrict $\mathscr{D}$ to a subclass of $\mathscr{D}$ may be categorized as follows:

§1. *Unbiasedness, Similarity*

The test functions of size α in this subclass are required to be similar in the sense that $\pi(\phi, \theta) \equiv \alpha$ for all $\theta \in \Theta_0$ and also unbiased in the sense that $\pi(\phi, \theta) \geq \alpha$ for all $\theta \in \Theta_1$.

§2. *Invariance*

The test functions in this subclass are required to be of certain special forms determined by the underlying group structure of the probability model $f(x \mid \theta)$ and the testing problem H versus K.

In addition, we may mention the following two additional criteria which are widely used as methods of constructing a test without any reference to the power function.

§3. *The Likelihood Ratio (LR) Principle*

A very general method of test construction.

§4. *Studentization*

An ad hoc method based on studentization of a suitable estimator of the parameter of interest.

Of course, these criteria are more or less mutually related. In fact, in most univariate cases and in some multivariate cases, these approaches often give the same test. However, the principle of unbiasedness or similarity, as mentioned in §1, is no longer effective in such incomplete models as the GMANOVA model, etc., because of the lack of the completeness of the models (see Chapter 5).

An excellent exposition dealing with the completeness of a model is found in Chapter 4 of Lehmann (1959). On the other hand, the invariance approach in §2 is often applicable to incomplete multivariate normal linear models. The likelihood ratio (LR) principle (§3) is also always applicable to any model whenever the form of the probability density is known, though it often fails to give an explicit form of the test. The philosophy behind this LR principle differs from that of the Neyman–Pearson theory, and so it might not be appropriate to compare the LRT (likelihood ratio test) to other optimal tests in terms of power. However, the LRT is always invariant (see Lehmann, 1959, p. 252 or Eaton, 1983, p. 263); hence, it is naturally included in the analysis through invariance. The ad hoc approach in §4 is often difficult to apply in the case of incomplete models because a good estimator is difficult to find there. In addition, the studentization of a good estimator does not necessarily produce a good test. Further, the tests thus obtained are mostly invariant. On the basis of these aspects, *in this book we adopt the invariance principle as our basic standpoint.*

For a general theory on invariance, Lehmann (1959), Ferguson (1967), or Eaton (1983) will serve as basic references. To describe the invariance principle in connection with tests of hypotheses, let

$$\mathscr{P}(\Theta) = \{P_\theta \,|\, dP_\theta/d\mu = f(x\,|\,\theta), \qquad f(x\,|\,\theta) \in \mathscr{F}(\Theta)\} \tag{2.19}$$

where $\mathscr{F}(\Theta)$ is given by (2.15), and let $\mathscr{G}$ be a group of bijective and bimeasurable transformations from $\mathscr{X}$ onto $\mathscr{X}$ such that

$$P_\theta \in \mathscr{P}(\Theta) \Rightarrow gP_\theta \in \mathscr{P}(\Theta) \qquad \text{for any} \quad g \in \mathscr{G} \tag{2.20}$$

where $gP_\theta(A) = P_\theta(g^{-1}(A))$ for a Borel set A of $\mathscr{X}$. When a given model $\mathscr{P}(\Theta)$ in (2.19) or $\mathscr{F}(\Theta)$ in (2.15) admits such a group $\mathscr{G}$, it is said to be invariant under the group $\mathscr{G}$ or, in other words, $\mathscr{G}$ preserves model $\mathscr{P}(\Theta)$. The relation (2.20) with the identifiability of $\mathscr{F}(\Theta)$ means there exists a parameter θ' in Θ such that $gP_\theta = P_{\theta'}$, and the correspondence between θ and θ' is designated by $\theta' = \bar{g}\theta$. Then $\bar{\mathscr{G}} = \{\bar{g}\,|\,g \in G\}$ forms a group as a homomorphic image of $\mathscr{G}$. In this situation, if the induced group $\bar{\mathscr{G}}$ leaves the testing problem (2.16) invariant in the sense that

$$\bar{g}(\Theta_0) = \Theta_0 \qquad \text{and} \quad \bar{g}(\Theta_1) = \Theta_1 \quad \text{for any} \quad \bar{g} \in \bar{\mathscr{G}}, \tag{2.21}$$

it is said that the testing problem is left invariant under the group $\mathscr{G}$. The condition (2.21) means that $\bar{\mathscr{G}}$ preserves the hypotheses or $\mathscr{G}$ preserves $\mathscr{P}(\Theta_i)$'s. Now when a testing problem is left invariant under a group $\mathscr{G}$, we require a test function $\phi(x)$ to be invariant;

$$\phi(gx) = \phi(x) \qquad \text{for all} \quad g \in \mathscr{G} \quad \text{and} \quad x \in \mathscr{X}, \tag{2.22}$$

and restrict our attention to the class of invariant tests, say $\mathscr{D}_{\mathrm{I}}$. The requirement (2.22) says that any point in the orbit of x defined by

$$\mathscr{G}x = \{gx \in \mathscr{X}\,|\,g \in \mathscr{G}\}$$

is considered equivalent to x for making a decision on the problem. That is, we reject the null hypothesis with the same probability $\phi(x)$ for any observation falling in $\mathscr{G}x$. Hence if $\phi(x_1) \neq \phi(x_2)$, x_1 and x_2 are on different orbits, or $\mathscr{G}x_1 \neq \mathscr{G}x_2$. A measurable function $T(x)$ from $\mathscr{X}$ into a measurable space $\mathscr{T}$ satisfying

(a) $T(gx) = T(x)$ for all $g \in \mathscr{G}$ and $x \in \mathscr{X}$, and
(b) $T(x_1) = T(x_2)$ implies $x_1 = gx_2$ for some $g \in \mathscr{G}$

is called a maximal invariant. Condition (a) says that a maximal invariant is constant on orbits, whereas condition (b) says that each orbit gets a different value, that is, T distinguishes orbits. Therefore an invariant test $\phi(x)$ is expressed as (Exercise 2.12)

$$\phi(x) = \psi(T(x)) \qquad \text{for some test } \psi \text{ on } \mathscr{T}. \tag{2.23}$$

Also a maximal invariant parameter $\tau(\theta)$ is defined to be a map τ from Θ into a space Υ such that

$$\tau(\bar{g}\theta) = \tau(\theta) \qquad \text{for all} \quad \theta \in \Theta \quad \text{and all} \quad \bar{g} \in \bar{\mathscr{G}} \qquad \text{and}$$

$$\tau(\theta_1) = \tau(\theta_2) \quad \text{implies} \quad \theta_1 = \bar{g}\theta_2$$

for some $\bar{g} \in \bar{\mathscr{G}}$. Under this situation, the following lemma is well known and easy to prove (see Lehmann, 1959, p. 220).

Lemma 2.10. *The distribution of a maximal invariant $T(x)$ under θ depends on θ only through a maximal invariant parameter $\tau(\theta)$. Hence for any $\phi \in \mathscr{D}_1$, the power function $\pi(\phi, \theta) = E_\theta[\phi(x)]$ depends on θ only through $\tau(\theta)$.*

With this lemma, the problem of testing H: $\theta \in \Theta_0$ versus K: $\theta \in \Theta_1$ in (2.16) is now reduced by invariance as follows. Let $\mathscr{D}_1^T$ be the class of tests on the range space $\mathscr{T}$ of $T(x)$, and let

$$\mathscr{P}^T = \{P_\tau^T \mid \tau = \tau(\theta), \qquad P_\tau^T = P_\theta \circ T^{-1}, \qquad P_\theta \in \mathscr{P}(\Theta)\} \tag{2.24}$$

be the class of probability distributions induced by $T = T(x)$ through $\mathscr{P}(\Theta)$ in (2.19). Then the problem is regarded as

$$\text{H}': \tau \in \tau(\Theta_0) \qquad \text{versus} \qquad \text{K}': \tau \in \tau(\Theta_1), \tag{2.25}$$

and in this reduced problem, we wish to maximize in some sense the power function

$$\tilde{\pi}(\psi, \tau) = E_\tau[\psi(T)] = E_\theta[\psi(T(x))] = \pi(\psi \circ T, \theta)$$

with respect to ψ in $\mathscr{D}_{\mathrm{I}}^T$. In this maximization, the (generalized) Neyman–Pearson lemma is applied to the class $\mathscr{P}^T$ in (2.24). A test ϕ^* in $\mathscr{D}_{\mathrm{I}}$ (or ψ^* in $\mathscr{D}_{\mathrm{I}}^T$ with $\phi^* = \psi^* \circ T$) is called a UMPI (uniformly most powerful invariant) test if for any $\phi \in \mathscr{D}_{\mathrm{I}}$

$$\pi(\phi^*, \theta) \geq \pi(\phi, \theta) \qquad \text{for all} \quad \theta \in \Theta_1,$$

and a test ϕ^* in $\mathscr{D}_{\mathrm{I}}$ (or ψ^* in $\mathscr{D}_{\mathrm{I}}^T$ with $\phi^* = \psi^* \circ T$) is called an LBI (locally best invariant) test if there exists an open neighborhood $\tilde{\Theta}_1$ of Θ_0 such that for any $\phi \in \mathscr{D}_{\mathrm{I}}$

$$\pi(\phi^*, \theta) \geq \pi(\phi, \theta) \qquad \text{for all} \quad \theta \in \tilde{\Theta}_1 - \Theta_0.$$

Of course, even in the restricted class $\mathscr{D}_{\mathrm{I}}$, there may not exist in general a UMPI test or an LBI test, although the latter exists in many multivariate testing problems under normality assumption.

To summarize, the following procedures are usually taken for analysis of an invariant testing problem:

(1) Finding a group which leaves the given problem invariant.

(2) Choosing a convenient maximal invariant under the group found in step (1).

(3) Deriving the distribution of the maximal invariant chosen in step (2).

(4) Deriving an optimal (UMPI, LBI, etc.) test based on the distribution of the maximal invariant.

(5) Deriving or approximating the null distribution of the optimal test derived.

In addition, one may also check the performance of the power function. For this purpose, one may consider the following points:

(6) Deriving the nonnull distribution of the test statistic and checking the behavior of the power function.

(7) Investigating whether the test is minimax or locally minimax. A test ϕ^* is called minimax in $\mathscr{D}$ with respect to Λ if

$$\min_{\theta \in \Lambda} \pi(\phi^*, \theta) = \sup_{\phi \in \mathscr{D}} \min_{\theta \in \Lambda} \pi(\phi, \theta),$$

where Λ is a suitable subset of Θ_1.

(8) Examining whether the power function is monotonically increasing as the parameter goes farther from the null hypothesis. This property is often called the monotonicity property of the power function.

(9) Checking whether the test is admissible in $\mathscr{D}_{\mathrm{I}}$ or more generally in $\mathscr{D}$. A test ϕ^* of size α in $\mathscr{D}_{\mathrm{I}}$ (or $\mathscr{D}$) is called admissible if there exists no test ϕ of size α in $\mathscr{D}_{\mathrm{I}}$ (or $\mathscr{D}$) such that

$$\pi(\phi, \theta) \geq \pi(\phi^*, \theta) \qquad \text{for all} \quad \theta \in \Theta_1$$

$$\pi(\phi, \theta) \leq \pi(\phi^*, \theta) \qquad \text{for all} \quad \theta \in \Theta_0$$

with strict inequality for at least one $\theta \in \Theta_1 \cup \Theta_0$.

(10) Investigating if the test is robust. In other words, one may question if, when the model is enlarged, the null distribution of the test remains the same and the optimality of the test is preserved. Those properties are often called the null robustness and the optimality robustness. We will return to these concepts of robustness in Chapter 3.

2.4. Distribution of a Maximal Invariant

As has been stated in Section 2.3, that when a testing problem is left-invariant under a group $\mathscr{G}$, it is a standard procedure to choose a convenient maximal invariant [step (2)] and then derive the distribution of the maximal invariant [step (3)]. However, sometimes maximal invariants are too complicated to treat analytically. A typical example is the GMANOVA problem. In addition, it is not always necessary to derive an explicit form of the distribution. Indeed, when we are interested in locally optimal tests, we simply need a local form of the distribution in the neighborhood of the null hypothesis. In such a case, and also in other cases, what is known as a representation of the probability ratio of a maximal invariant is very useful. There are several versions of the representation depending upon conditions on the group leaving a testing problem invariant and the sample space acted upon by the group, but all of them give the same expression:

$$\text{(2.26)} \qquad R(T(x)) \equiv \frac{dP^T_{\theta_1}}{dP^T_{\theta_0}}(T(x)) = \frac{\displaystyle\int_{\mathscr{G}} f(gx \mid \theta_1)\chi(g)\nu(dg)}{\displaystyle\int_{\mathscr{G}} f(gx \mid \theta_0)\chi(g)\nu(dg)}.$$

Here $T(x)$ is a maximal invariant, P^T_θ is the distribution of $T(x)$ under θ, $f(x \mid \theta)$ is a density with respect to a relatively invariant measure μ, $\mathscr{G}$ is a locally compact group, ν is a left-invariant measure on $\mathscr{G}$, and $\chi(g)$ is a multiplier function of μ (see Section 2.1). A few sets of precise conditions for which (2.26) holds are stated below. The representation (2.26) says that the Radon–

Nikodym derivative of the distribution of a maximal invariant under θ_1 with respect to that under θ_0, when evaluated at $T = T(x)$, is given by the ratio of the integrals of the densities under θ_1 and θ_0 over the group $\mathscr{G}$. In testing problems, θ_1 and θ_0 are respectively chosen from the alternative hypothesis space Θ_1 and the null hypothesis space Θ_0 so that (2.26) represents the likelihood ratio of $T(x)$ under θ_1 and θ_0. Hence, if the null distribution $P_{\theta_0}^T$ of a maximal invariant T is the same for all $\theta_0 \in \Theta_0$, which is true when Θ_0 is simple, by the Neyman–Pearson lemma, a test with critical region $R(T(x)) > c$ is a MPI (most powerful invariant) test for testing Θ_0 versus the fixed θ_1 in Θ_1. If the test happens to be independent of θ_1, it is UMPI for testing Θ_0 versus Θ_1. In case this does not happen and we are interested in a locally optimal test, the ratio in (2.26) is expanded in the neighborhood of Θ_0, and the integral is evaluated locally. In these procedures, it is not at all necessary to choose an explicit form of a maximal invariant $T(x)$, and step (3) can be skipped altogether.

It was Stein (1966) who first gave the expression (2.26), but he did not explicitly state the conditions for which it is valid. Stein's representation was applied by Giri (1968) and Schwartz (1967a), among others. In an unpublished paper, Schwartz (1967) gave a set of conditions for the validity of (2.26), but his conditions are rather complicated. His result is introduced in Farrell (1976). Wijsman (1967, 1969) took a differential-geometric approach and gave a sufficient condition for (2.26) by using the concept of Cartan $\mathscr{G}$ space. Koehn (1970) generalized some results of Wijsman (1967), whereas Bondar (1976) considered conditions for (2.26) through topological argument. Recently, by taking Bourbaki's approach (1963, 1966) and by using the notion of a quotient measure, Andersson (1982) established some results concerning (2.26) in terms of the proper action of a group. Wijsman (1985) studied the properness of the action in several applied multivariate problems. Furthermore, Wijsman (1986) also investigated the global cross-section for factorization of measures and applied it to the representation of the distribution of a maximal invariant.

I. Andersson's Approach

On the basis of Bourbaki's approach, Andersson (1982), by using quotient measures, obtains the representation of densities of maximal invariants. His framework is relatively simpler than Wijsman's (1967), as will be seen below. Let $\mathscr{G}$ be a locally compact sigma-compact Hausdorff topological group and let $\mathscr{X}$ be a locally compact sigma-compact Hausdorff topological space. Suppose $\mathscr{G}$ acts topologically on $\mathscr{X}$. A left-invariant measure on $\mathscr{G}$ is denoted by ν, and let $\mathscr{X}/\mathscr{G}$ be the quotient space $\mathscr{X}/\mathscr{G} \equiv \mathscr{Q}$ with the quotient topology. In

Andersson (1982), the natural projection π from $\mathscr{X}$ onto the quotient space $\mathscr{Q}$ is taken as a maximal invariant. In fact, $\pi(x) = \mathscr{G}x$ for $x \in \mathscr{X}$ is a maximal invariant. Further, let $f_i(x)$ $(i = 1, 2)$ be two probability densities on $\mathscr{X}$ with respect to a relatively invariant measure μ with multiplier χ. Under this setup a sufficient condition for the representation (2.26) to hold with $T = \pi$ is provided by the notion of a proper action:

Definition 2.9. Consider the map K of $\mathscr{G} \times \mathscr{X}$ into $\mathscr{X} \times \mathscr{X}$ given by $K(g, x) = (gx, x)$. Then the action of $\mathscr{G}$ on $\mathscr{X}$ is said to be proper if $K^{-1}(C)$ is compact for each compact subset $C \subset \mathscr{X} \times \mathscr{X}$. The space $\mathscr{X}$ is called a proper $\mathscr{G}$-space if the action of $\mathscr{G}$ is proper.

Theorem 2.4. (*Andersson*, 1982). *Suppose $\mathscr{G}$ acts properly on $\mathscr{X}$. Then the probability ratio of the maximal invariant π under f_i $(i = 1, 2)$ is given by*

$$(2.27) \qquad R(\pi(x)) \equiv \frac{dP_2^\pi}{dP_1^\pi}(\pi(x)) = \frac{\displaystyle\int_{\mathscr{G}} f_2(gx)\chi(g)\nu(dg)}{\displaystyle\int_{\mathscr{G}} f_1(gx)\chi(g)\nu(dg)} \quad \text{a.e.} \quad (P_1^\pi).$$

A sketch of the proof of this theorem is provided in Appendix 2 of this chapter.

Definition 2.1 of the properness of the action of $\mathscr{G}$ is not easy to handle as it stands. To give a set of equivalent conditions, define for $A, B \subset \mathscr{X}$,

$$(2.28) \qquad ((A, B)) = \{g \in \mathscr{G} \mid gA \cap B \neq \phi\}.$$

Definition 2.10. $\mathscr{X}$ is called a Cartan $\mathscr{G}$-space if for each $x \in \mathscr{X}$, there exists a neighborhood V of x such that $((V, V))$ has compact closure.

Lemma 2.11. *Under the present setup, the following conditions are equivalent.*

(1) *$\mathscr{G}$ acts properly on $\mathscr{X}$.*

(2) *For any x and y in $\mathscr{X}$, there exist neighborhoods V_x and V_y of x and y respectively, such that $((V_x, V_y))$ has compact closure.*

(3) *$\mathscr{X}$ is a Cartan $\mathscr{G}$-space and $\mathscr{X}/\mathscr{G}$ is Hausdorff.*

(4) *For $K \subset \mathscr{X}$ compact, $((K, K))$ has compact closure.*

Proof. See Palais (1961) and Bourbaki (1963, 1966).

As a matter of fact, the sigma compactness of $\mathscr{G}$ and $\mathscr{X}$ assumed in our setup is not necessary in order to obtain the result in Lemma 2.11. It should also be remarked that if A and B are compact in $\mathscr{X}$, $((A, B))$ is closed. The following result is useful in applications.

Lemma 2.12. (*Palais*, 1961, 1.3.3). *Let $\mathscr{Y}$ be a locally compact space acted upon topologically by a locally compact group $\mathscr{G}$. Then if $\mathscr{X}$ is a proper* (*Cartan*) *$\mathscr{G}$-space, so is $\mathscr{X} \times \mathscr{Y}$.*

Wijsman (1985) develops a useful tool for proving the action of $\mathscr{G}$ on $\mathscr{X}$ to be proper.

§II. *Wijsman's Approach*

Let $\mathscr{X}$ be a nonempty open subset of R^n, $\mathscr{G}$ a Lie subgroup of the general linear group GL(n) acting linearly on $\mathscr{X}$, and ν the Lebesgue measure on $\mathscr{X}$.

Theorem 2.5. (*Wijsman*, 1967). *Suppose $\mathscr{X}$ is a Cartan $\mathscr{G}$-space. Then* (2.26) *holds with $\chi(g) = |g'g|^{1/2}$ for $g \in \mathscr{G}$.*

The proof requires some knowledge of differential geometry. Later on, Wijsman (1969) extended the result to the case where $\mathscr{G}$ is a Lie subgroup of the general affine linear group.

Although the setup for $\mathscr{X}$ and $\mathscr{G}$ in Theorem 2.5 is more restrictive than that in Theorem 2.4, as is shown in Lemma 2.11, the condition for $\mathscr{X}$ to be a Cartan $\mathscr{G}$-space is weaker and easier to check than the condition for $\mathscr{X}$ to be a proper $\mathscr{G}$-space.

To state a condition for $\mathscr{X}$ to be a Cartan $\mathscr{G}$-space, let $\mathscr{G}$ be a locally compact group and $\mathscr{X}$ a completely regular Hausdorff space. When $\mathscr{G}$ acts freely on $\mathscr{X}$ (i.e., $g \neq e$ implies $gx \neq x$ for all $x \in \mathscr{X}$), $\mathscr{X}$ is said to be a $\mathscr{G}$ principal bundle. If $\mathscr{X}$ is a $\mathscr{G}$ principal bundle, for each $(x_1, x_2) \in \mathscr{R}$ there is a unique element $f(x_1, x_2) \in \mathscr{G}$ such that $x_2 = f(x_1, x_2)x_1$, where

$$\mathscr{R} = \{(x_1, x_2) \in \mathscr{X} \times \mathscr{X} \mid x_1 = gx_2 \quad \text{for some} \quad g \in \mathscr{G}\}.$$

Hence if f is continuous from $\mathscr{R}$ into $\mathscr{G}$, $\mathscr{X}$ is called a Cartan principal bundle.

Lemma 2.13. (*Palais*, 1961, 1.1.3). *A $\mathscr{G}$-principal bundle $\mathscr{X}$ is Cartan principal bundle if and only if it, is a Cartan $\mathscr{G}$-space.*

Example 2.2. (MANOVA problem). Let $X\colon n \times p \sim N(\begin{pmatrix}\Theta\\0\end{pmatrix}, I_n \otimes \Sigma)$, where $X = (X_i)$ with $X_i\colon n_i \times p$ $(i = 1, 2, 3)$, $\Theta = (\Theta_j)\colon (n_1 + n_2) \times p$ with $\Theta_j\colon n_j \times p$, $(j = 1, 2)$, and $n_1 + n_2 + n_3 = n$ with $n_3 \geq p$. The MANOVA problem deals with $H\colon \Theta_2 = 0$, and the group $\mathscr{G} \equiv \mathcal{O}(n_1) \times \mathcal{O}(n_2) \times \mathcal{O}(n_3) \times \mathrm{GL}(p) \times R^{n_1 p}$ acts on the space of X by

$$(X_1, X_2, X_3) \to (\Gamma_1 X_1 A' + F, \Gamma_2 X_2 A', \Gamma_3 X_3 A'),$$

where $(\Gamma_1, \Gamma_2, \Gamma_3, A, F) \in \mathscr{G}$. It follows that a maximal invariant is a function of X_2 and X_3. Hence letting

$$\mathscr{X}_i = \{x \in R^{n_i \times p} \,|\, \mathrm{rank}(x) = \min(n_i, p)\} \qquad (i = 2, 3),$$

it suffices to show that $\mathscr{X}_2 \times \mathscr{X}_3$ is a Cartan $\mathcal{O}(n_2) \times \mathcal{O}(n_3) \times \mathrm{GL}(p)$ space because $R^{n_2 p} \times R^{n_3 p} - \mathscr{X}_2 \times \mathscr{X}_3$ has measure 0. Further, since $\mathcal{O}(n_2) \times \mathcal{O}(n_3)$ is compact, by Definition 2.10, it suffices to show that $\mathscr{X}_2 \times \mathscr{X}_3$ is a Cartan $\mathrm{GL}(p)$ space. But by Lemmas 2.12 and 2.13, it suffices to show that $X_3 A' = X_3$ for $X_3 \in \mathscr{X}_3$ implies $A = I_p$, which follows from the definition of $\mathscr{X}_3$ and $n_3 \geq p$. Therefore, by Theorem 2.5, the probability ratio of a maximal invariant is given by (2.26), where $\mathscr{G} = \mathcal{O}(n_2) \times \mathcal{O}(n_3) \times \mathrm{GL}(p)$, $f(X_2, X_3 \,|\, \theta)$ is the density of (X_2, X_3), $\theta_0 = (0, \Sigma)$, $\theta_1 = (\Theta_2, \Sigma)$, and $\chi(g) = |AA'|^{(n_2 + n_3)/2}$ is the inverse of Jacobian $(X_2, X_3) \to g(X_2, X_3)$ for $g \in \mathscr{G}$. As mentioned in Section 2.1, the left-invariant measure $\nu(dg)$ on $\mathscr{G}$ is given by the products of two invariant probability measures on $\mathcal{O}(n_2)$ and $\mathcal{O}(n_3)$ and the left-invariant measure $|A'A|^{-p/2}\, dA$ on $\mathrm{GL}(p)$.

In a series of papers, Wijsman (1967, 1978, 1985, 1986) demonstrated the verification of the conditions of Cartan $\mathscr{G}$-space and proper actions for a wide variety of multivariate testing problems. The verification of the conditions for the testing problems considered in this book is covered in these papers. The details of the verifications, which are similar to the above, appear as exercises in Chapter 3.

Appendix to Chapter 2, Section 4

Description of Theorem 2.4. In this appendix we shall describe a recent result due to Andersson (1982) concerning quotient measures and the representation of densities of maximal invariants. In what follows, the notation and terminology given in Nachbin (1965) will be used. Let $\mathscr{G}$ be a locally compact σ-compact topological group which acts topologically on the locally compact σ-compact space $\mathscr{X}$. A left (right) Haar measure on $\mathscr{G}$ is denoted by $\nu_1(\nu_r)$, and the right-hand modulus of $\mathscr{G}$ is Δ_r, so $\Delta_r \nu_r = \nu_1$. The natural projection π from $\mathscr{X}$ to the quotient space $\mathscr{X}/\mathscr{G} = \mathscr{Q}$ is a convenient and natural choice for a maximal invariant under the action of $\mathscr{G}$ on $\mathscr{X}$. All measures on $\mathscr{X}$ and $\mathscr{Q}$ will be Radon measures.

Let μ be a measure on $\mathscr{X}$ that is relatively invariant with multiplier Δ_r^{-1}, that is,

$$\int_{\mathscr{X}} f(g^{-1}x)\mu(dx) = \Delta_r^{-1}(g) \int_{\mathscr{X}} f(x)\mu(dx)$$

for all μ-integrable f. Ignoring questions of existence of integrals for a moment, consider

$$\tilde{f}(x) \equiv \int_{\mathscr{G}} f(gx)v_r(dg) \tag{A.2.1}$$

and note that $\tilde{f}(x) = \tilde{f}(hx)$, $h \in \mathscr{G}$, so $\tilde{f}$ is invariant. Thus, we can write $\tilde{f}(x) = \tilde{f}(\pi(x))$, where $\tilde{f}$ is defined on $\mathscr{Q}$. If α is a measure on $\mathscr{Q}$, we can then integrate $\tilde{f}$ over $\mathscr{Q}$. This integration will be denoted by

$$J_\alpha(f) \equiv \int_{\mathscr{Q}} \left(\int_{\mathscr{G}} f(gx)v_r(dg) \right) d\alpha. \tag{A.2.2}$$

An easy calculation shows that $J_\alpha(f)$ is relatively left-invariant with multiplier Δ_r^{-1}, that is, $J_\alpha(g_1 f) = \Delta_r^{-1}(g_1)J_\alpha(f)$. A question treated in Andersson (1978) is the following: Given μ on $\mathscr{X}$, which is relatively invariant with multiplier Δ_r^{-1}, under what conditions will there exist an α on $\mathscr{Q}$ so that

$$J_\alpha(f) = \int_{\mathscr{X}} f(x)\mu(dx) \tag{A.2.3}$$

for all μ-integrable f? A sufficient condition for the representation (A.2.3) to hold is provided by the notion of a proper action (see Definition 2.9).

Theorem A.2.1. (*See Andersson*, 1982). *Suppose the action of $\mathscr{G}$ on $\mathscr{X}$ is proper. If μ is a relatively left-invariant measure on $\mathscr{X}$ with multiplier Δ_r^{-1}, then there exists a measure α on $\mathscr{Q}$ such that*

$$\int_{\mathscr{X}} f(x)\mu(dx) = \int_{\mathscr{Q}} \left(\int_{\mathscr{G}} f(gx)v_r(dg) \right) d\alpha \tag{A.2.4}$$

for all μ-integrable f.

For the remainder of this appendix, it is assumed that $\mathscr{G}$ acts properly on $\mathscr{X}$. In some situations, one has a measure μ_0 on $\mathscr{X}$ which is relatively invariant with a multiplier χ_0, that is

$$\int f(g^{-1}x)\mu_0(dx) = \chi_0(g) \int f(x)\mu_0(dx)$$

for all integrable f and $g \in \mathscr{G}$. But a representation of the form (A.2.4) is still desired. To obtain such a result, the measure μ_0 needs to be modified. It is asserted in Andersson (1982) that there exists a positive continuous function η_0 on $\mathscr{X}$ such that

$$\eta_0(gx) = \Delta_r(g)\chi_0(g)\eta_0(x) \tag{A.2.5}$$

for $x \in \mathscr{X}$ and $g \in \mathscr{G}$. Setting $\mu = \eta_0^{-1}\mu_0$, it is easily verified that μ is relatively invariant with multiplier Δ_r^{-1}. Thus, applying the above theorem gives

$$\begin{aligned}\int_{\mathscr{X}} f(x)\mu_0(dx) &= \int_{\mathscr{X}} f(x)\eta_0(x)\mu(dx)\\ &= \int_{\mathscr{Q}}\left(\int_{\mathscr{G}} f(gx)\eta_0(gx)\nu_r(dg)\right)d\alpha\\ &= \int_{\mathscr{Q}}\left(\eta_0(x)\int_{\mathscr{G}} f(gx)\chi_0(g)\nu_1(dg)\right)d\alpha.\end{aligned} \tag{A.2.6}$$

where $\nu_1 = \Delta_r\nu_r$ is a left Haar measure on $\mathscr{G}$.

Following Andersson (1982), we will now apply (A.2.6) to find the density function of a maximal invariant. Let μ_0 be relatively invariant with multiplier χ_0 and suppose that the random variable $X \in \mathscr{X}$ has a density f_0 with respect to μ_0. The random variable $Y \equiv \pi(X) \in \mathscr{Q}$ is a maximal invariant. In the notation of (A.2.6), the claim is that the density of Y with respect to the measure α is ρ_0, where

$$\rho_0(\pi((x)) = \eta_0(x)\int_{\mathscr{G}} f_0(gx)\chi_0(g)\nu_1(dg). \tag{A.2.7}$$

Given (A.2.6), the verification of (A.2.7) is identical to the case when $\mathscr{G}$ is compact (see Stein, 1966, or Eaton, 1983). To verify that ρ_0 is the density of Y, it suffices to show that

$$Ek(Y) = \int_{\mathscr{Q}} k(y)\rho_0(y)\alpha(dy)$$

for suitably many functions k on $\mathscr{G}$. But

$$\begin{aligned}Ek(Y) = Ek(\pi(x)) &= \int k(\pi(x))f_0(x)\mu_0(dx)\\ &= \int_{\mathscr{Q}} \eta_0(x)k(\pi(x))\int_{\mathscr{G}} f_0(gx)\chi_0(g)\nu_1(dg)\,d\alpha\\ &= \int_{\mathscr{Q}} k(y)\rho_0(y)\alpha(dy),\end{aligned}$$

with the last equality following from the definition of ρ_0.

Our application of (A.2.7) concerns the ratio of two densities of a maximal invariant. If f_0 and f_1 are two possible densities of X, and ρ_0 and ρ_1 are the two induced densities of Y, then from (A.2.7) the ratio $r(y) = \rho_1(y)/\rho_0(y)$ is given by

$$r(y) = \frac{\int_{\mathscr{G}} f_1(gx)\chi_0(g)v_1(dg)}{\int_{\mathscr{G}} f_0(gx)\chi_0(g)v_1(dg)} \tag{A.2.8}$$

as long as the denominator is positive.

Exercises

2.1. Find the inverse of $(A, b) \in \mathrm{AL}(p)$ based on (2.1).

2.2. Show that $\mathrm{GL}(p)$, $\mathrm{GU}(p)$ and $\mathrm{AL}(p)$ are locally compact and sigma-compact, and that $\mathscr{O}(p)$ is compact.

2.3. Show that the groups in Exercise 2.2 above are separable metric spaces.

2.4. Prove (2.4).

2.5. Using Theorem 2.2, show that $I(fa) = \Delta(a)I(f)$.

2.6. Show that a right-modular function Δ is a homomorphism.

2.7. Show that a multiplier χ is a homomorphism.

2.8. Find a right-invariant measure on $\mathrm{AL}(p)$.

2.9. Show that $\mathscr{K}$ in (2.5) is a locally compact group, and derive the left-invariant measure μ on it.

2.10. Prove Lemma 2.3.

2.11. By using Lemma 2.4, prove Lemma 2.5.
[Hint: $\int \operatorname{tr} A\Gamma B\Gamma' v(d\Gamma) = \Sigma\Sigma\Sigma\Sigma a_{ij} b_{kl} \int \gamma_{jl}\gamma_{il} v(d\Gamma)$.]

2.12. Prove that an invariant function is a function of a maximal invariant.

2.13. Prove Lemma 2.6.

2.14. Show that T in (2.6) satisfies (i) and (ii) of Definition 2.6.

2.15. Prove that π_x in (2.7) is surjective (onto) and $\mathscr{G}x = \mathscr{X}$.

2.16. Show that if $\mathscr{G}$ is free with $(\mathscr{G}, \mathscr{X}, T)$ topological, the invariant measure defined by (2.10) is the unique invariant measure on $(\mathscr{X}, \tilde{\mathscr{D}})$.

2.17. Prove Lemma 2.7.

2.18. Prove Lemma 2.8.

Chapter 3 General Approach to the Robustness of Tests

3.1. Null, Nonnull, and Optimality Robustness

The purpose of this chapter is to introduce some concepts of the robustness of tests and to develop a general theory through which the robustness properties of a given test can be verified. As mentioned in Chapter 1, a multivariate parametric testing problem is often considered under a multivariate normal model. Then, depending on the testing problem, some tests are derived or proposed. The critical values of the test statistics are computed under the null hypothesis in the assumed model, whereas the powers of the tests are computed under the nonnull hypothesis in the assumed model. However, the assumed model may not represent or even approximate the process of the generation of the given data, and even if it does, the critical values or the null distributions and the powers or the nonnull distributions of the test statistics are likely to be sensitive to the deviations from the assumed model. Hence, in the presence of some uncertainty in the validity of the assumed model, it is important to study the stability or robustness of the critical values of the tests and also the optimality properties of the tests in terms of their power functions, if any, against departures from the assumed model. The robustness of a test may thus be studied from two aspects: (1) robustness under a null hypothesis and (2) robustness under an alternative hypothesis. By (1), it is usually meant that the critical value of a size α test computed under an assumed model is stable against a departure from the model, whereas, by (2), it is generally understood that the optimality properties a test may enjoy under the assumed model are stable against a departure from the model. Our concepts of the robustness of a test are similar to these concepts but more specific and stronger. To define the concepts, let

$$\mathscr{F}(\Theta) = \{f(\cdot \mid \theta) \mid \theta \in \Theta\} \tag{3.1}$$

be the class of the densities of an assumed parameteric model and let

$$\text{H}: \theta \in \Theta_0 \qquad \text{versus} \qquad \text{K}: \theta \in \Theta_1 \tag{3.2}$$

be a testing problem under consideration, where $f(\cdot \mid \theta)$ is the density of a probability measure P_θ on a sample (measurable) space $(\mathscr{X}, \mathscr{B})$ with respect to a sigma-finite measure μ, and Θ is a parameter space with $\Theta = \Theta_0 \cup \Theta_1$ and $\Theta_0 \cap \Theta_1 = \phi$. Here the form of the density $f(\cdot \mid \theta)$ is assumed to be known, and Θ may be identified with a subset of an Euclidean space. It should be remarked that the model may be stated by the class of probability measures

$$\mathscr{P}(\Theta) = \{P_\theta \mid \theta \in \Theta\} \tag{3.3}$$

instead of the densities $f(\cdot \mid \theta)$ if the dominating measure μ does not exist. The framework here is broad enough to cover almost all applications. Next let

$$\mathscr{H}(\Theta) = \{h(\cdot \mid \theta) \mid \theta \in \Theta, h \in \mathscr{H}\} \tag{3.4}$$

be an enlarged class of densities with respect to μ, where $\mathscr{H}$ is a class of functions $h(\cdot \mid \cdot)$ defined on $\mathscr{X} \times \Theta$, such that $h(\cdot \mid \theta)$ is a density for each $\theta \in \Theta$ and $\mathscr{H}$ contains $f(\cdot \mid \cdot)$. Hence $\mathscr{F}(\Theta) \subset \mathscr{H}(\Theta)$, and each $h(\cdot \mid \theta)$ in $\mathscr{H}(\Theta)$ has the same parameter space Θ as $f(\cdot \mid \theta)$, so that the testing problem (3.2) can be considered under each density $h(\cdot \mid \theta)$ in $\mathscr{H}(\Theta)$. Because in applications a test function is described by a test statistic, our concept of the robustness of a test under the null hypothesis is defined in terms of test statistics as follows.

Definition 3.1. A (test) statistic $T = T(x)$ for the problem (3.2) considered under the model (3.1) is said to be null-robust against the departure from $\mathscr{F}(\Theta)$ toward $\mathscr{H}(\Theta)$ if for each null point $\theta \in \Theta_0$, the (null) distribution of T under $f(x \mid \theta)$ is exactly the same as that under any $h(x \mid \theta)$ in $\mathscr{H}(\Theta)$.

The null robustness in this definition is a finite sample property, and it guarantees the complete stability or invariance of critical values of the test statistic T in the class $\mathscr{H}(\Theta_0)$. Some conditions for a test statistic to be null-robust are given in Section 3. The definition is extended to a general statistic T in an obvious way.

Example 3.1. (See Example 1.1 of Chapter 1.) Let $x = (x_1, \ldots, x_n)'$ be an observation from $N(\mu \underset{\sim}{1}, \sigma^2 I_n)$ where $(\mu, \sigma^2) \in \Theta = R \times R_+$ and $\underset{\sim}{1} = (1, \ldots, 1)' \in R^n$, and consider the testing problem H: $\mu = 0$ versus K: $\mu \neq 0$. Then $\mathscr{X} = R^n$, μ is the Lebesgue measure on $\mathscr{X}$, $\Theta_0 = \{0\} \times R_+$, and $\Theta_1 = (R - \{0\}) \times R_+$, and $\mathscr{F}(\Theta) = \{f(x \mid \theta) \mid \theta \in \Theta\}$ is the class of the normal densities of $N(\mu \underset{\sim}{1}, \sigma^2 I_n)$. Now let us consider the null robustness of the t-test statistic

$$t = t(x) = \sqrt{n}\,\bar{x}/s,$$

where $\bar{x} = \Sigma x_i/n$ and $s^2 = \Sigma(x_i - \bar{x})^2/(n-1)$. An important class of densities into which we may look for $\mathscr{H}(\Theta)$ in (3.4) is the class of all the normal mixtures

$$h(x \mid \theta) = \int_0^\infty f(x \mid (\mu, a\sigma^2))\, dG(a)$$

where G is a distribution function on R_+. Such a choice of $\mathscr{H}(\Theta)$ is easily justified because this class is a naturally enlarged class of the normal class $\mathscr{F}(\Theta)$, which preserves the invariance structure of the original problem (see Chapters 1 and 2). Then, as will be shown in Section 3, the t-statistic is null-robust (against the departure from $\mathscr{F}(\Theta)$ toward $\mathscr{H}(\Theta)$).

Next, we shall define two concepts of robustness under the alternative hypothesis, namely what we call nonnull robustness and optimality robustness. The concept of the nonnull robustness of a test is similar to that of the null robustness and is defined as follows.

Definition 3.2. A test statistic $T = T(x)$ for problem (3.2) in model (3.1) is said to be nonnull-robust against the departure from $\mathscr{F}(\Theta)$ toward $\mathscr{H}(\Theta)$ if for each nonnull point $\theta \in \Theta_1$ the (nonnull) distribution of T under $f(x \mid \theta)$ is exactly the same as that under any $h(x \mid \theta)$ in $\mathscr{H}(\Theta)$.

Generally speaking, the nonnull robustness is hard to hold. Sufficient conditions for a test to be nonnull-robust are given in Section 3.4. There invariant tests for intraclass correlation, heteroscedasticity, serial correlation, independence, and sphericity are shown to be nonnull-robust. Some invariant tests for detection of outliers are also nonnull-robust (see Chapter 7).

The optimality robustness, on the other hand, is the stability of optimality properties concerning the behavior of power functions such as UMP (uniformly most powerful), UMPI (UMP invariant), and LBI (locally best invariant) properties, the monotonicity of power functions, admissibility, minimaxity, etc. Formally optimality-robustness is defined as

Definition 3.3. A test for problem (3.2) in model (3.1) is said to be optimality-robust against the departure from $\mathscr{F}(\Theta)$ toward $\mathscr{H}(\Theta)$ if an optimality property of the power function of the test under $\mathscr{F}(\Theta)$ holds exactly under $\mathscr{H}(\Theta)$.

Clearly, nonnull robustness implies optimality robustness whatever optimality of power may be concerned. Hence nonnull-robust tests for independence, heteroscedasticity, serial correlation, sphericity, etc., to be established in Section 4 are optimality-robust. The UMP and UMPI properties of the t-test in Example 3.1 are shown to be robust in Section 3.5 and in Chapter 4. The UMPI property of the F-test in the ANOVA problem and the UMPI property of the test for independence based on the multiple correlation are shown to be robust in Chapters 5 and 6, respectively. The LBI tests in the MANOVA and GMANOVA problems are shown to be optimality-robust in Chapter 5, whereas the LBI test for independence in general is shown to be optimality-robust in Chapter 6. The optimality robustness of LBI tests for detection of outliers will be treated in Chapter 7.

This chapter is organized as follows. An outline of most standard parametric testing problems under the assumption of multinormality of data is presented in Section 3.2. In Section 3.3, a necessary and sufficient condition for the null robustness of a statistic, which is not necessarily a test statistic, will be given. In Section 3.4, in the context of an invariant testing problem sufficient conditions for the nonnull robustness of a maximal invariant statistic will be established. Hence if the condition is satisfied, all invariant tests are nonnull-robust, since an invariant test is a function of a maximal invariant. In Section 3.5, a unified approach to the optimality-robustness will be developed.

3.2. Outline of Testing Problems under Normality

Canonically reduced forms of many testing problems under normality often encountered in applications concern hypotheses on the parameters (μ, Φ, Σ) of the normal distribution

$$(3.5) \qquad N_{np}(\mu, \Phi \otimes \Sigma), \qquad \mu \in R^{n \times p}, \qquad \Phi \in \mathscr{S}(n), \qquad \Sigma \in \mathscr{S}(p)$$

with a data matrix $X: n \times p$, where the notation in (3.5) is defined in Section 1.1, Chapter 1. In this section, some important problems that we will treat in this book are introduced.

A. GMANOVA, MANOVA and ANOVA Problems

The GMANOVA (general multivariate analysis of variance) problem, which we shall consider in a nonnormal setup in Chapter 5, is the problem of testing a general linear hypothesis of the form

(3.6) $$\text{H}: R_3 B R_4 = R_0 \qquad \text{versus} \qquad \text{K}: R_3 B R_4 \neq R_0$$

in the generalized multivariate regression model or growth curve model

(3.7) $$Y = R_1 B R_2 + E, \qquad E \sim N(0, I_n \otimes \Omega),$$

where $R_1: n \times (n_1 + n_2)$ of rank $n_1 + n_2$, $R_2: (p_1 + p_2) \times p$ of rank $p_1 + p_2$, $R_3: n_1 \times (n_1 + n_2)$ of rank n_1, and $R_4: (p_1 + p_2) \times p_2$ of rank p_2 are known matrices, $n = n_1 + n_2 + n_3$, $p = p_1 + p_2 + p_3$, and $B: (n_1 + n_2) \times (p_1 + p_2)$ and $\Omega \in \mathscr{S}(p)$ are unknown parameters. In this rather simple formulation, many hypothesis testing problems commonly treated in practice are included. For example, when $R_2 = I$ and $R_4 = I$, the problem is the well-known MANOVA problem, which alone covers a lot of interesting problems, such as the problem of testing on means, the equality of mean vectors, regression coefficients, and so on. The MANOVA problem with $p = 1$ is the ANOVA problem dealing with testing on regression coefficients. On the other hand, some important problems in applications that are not in the framework of the MANOVA problem are included in the GMANOVA problem. Examples are the problems of testing in growth curves used in biometrics and econometrics, testing on means with prior information, testing the equality of means in a classification model, etc. For details, the reader is referred to the recent books by Kariya (1985) and von Rosen (1985) on the GMANOVA problem and its extended problems.

The GMANOVA problem also entails some features of theoretical interest that the MANOVA problem does not provide. The features are in fact created by the presence of the matrices R_2 and R_4 in (3.7) and (3.6), respectively. One of the features is that the information provided by R_2 and R_4 produces an ancillary statistic, a statistic which is a part of a minimal sufficient statistic but whose marginal distribution does not depend on unknown parameters. It turns out that the likelihood ratio test (LRT), which ignores the ancillary statistic, is inadmissible (Marden, 1983). This is rather surprising because no other such example has ever been found in practical problems involving normal models. Below, we shall give canonical forms of the GMANOVA, MANOVA, and ANOVA problems for our future use. These canonical forms are derived in Chapter 5.

(A1) GMANOVA:

A canonical form of the GMANOVA model in (3.7) is given by

$$\text{(3.8)}\qquad \begin{matrix} & p_1 & p_2 & p_3 & \\ & \left(\begin{matrix} X_{11} \\ X_{21} \\ X_{31} \end{matrix}\right. & \begin{matrix} X_{12} \\ X_{22} \\ X_{32} \end{matrix} & \left.\begin{matrix} X_{13} \\ X_{23} \\ X_{33} \end{matrix}\right) & \begin{matrix} n_1 \\ n_2 \\ n_3 \end{matrix} \end{matrix} \sim N_{np}\left(\begin{pmatrix} \theta_{11} & \theta_{12} & 0 \\ \theta_{21} & \theta_{22} & 0 \\ 0 & 0 & 0 \end{pmatrix}, \quad I_n \otimes \Sigma \right),$$

in which the problem is to test

$$\text{(3.9)}\qquad \text{H: } \theta_{12} = 0 \quad \text{versus} \quad \text{K: } \theta_{12} \neq 0.$$

(See Gleser and Olkin, 1970, and Kariya, 1978, 1985). Clearly this model is a special case of (3.5). In Chapter 5, a group leaving this problem invariant is given and the problem is analyzed in a wider framework.

(A2) MANOVA:

Setting $p_1 = p_3 = 0$ in the canonical form of the GMANOVA model (3.8) yields a canonical form of the MANOVA model, which we write anew as

$$\text{(3.10)}\qquad X \equiv \begin{matrix} p & \\ \begin{pmatrix} X_1 \\ X_2 \\ X_3 \end{pmatrix} & \begin{matrix} n_1 \\ n_2 \\ n_3 \end{matrix} \end{matrix} \sim N_{np}\left(\begin{pmatrix} \mu_1 \\ \mu_2 \\ 0 \end{pmatrix}, \quad I_n \otimes \Sigma \right),$$

and the problem is to test H: $\mu_1 = 0$ versus K: $\mu_1 \neq 0$, where $n_3 \geq p$ is assumed. This problem is easily shown to be invariant under the group

$$\text{(3.11)}\qquad \mathscr{G} = \mathscr{O}(n_1) \times \mathscr{O}(n_2) \times \mathscr{O}(n_3) \times \text{GL}(p) \times R^{n_2 \times p}$$

acting on X by

$$\text{(3.12)}\qquad X \to \begin{pmatrix} Q_1 & 0 & 0 \\ 0 & Q_2 & 0 \\ 0 & 0 & Q_3 \end{pmatrix} XC' + \begin{pmatrix} 0 \\ F \\ 0 \end{pmatrix}$$

where $(Q_1, Q_2, Q_3, C, F) \in \mathscr{G}$. It can be shown that a maximal invariant is the set of the ordered nonzero latent roots of $(X_1'X_1)(X_3'X_3)^{-1}$ or, equivalently, the set of the ordered nonzero latent roots of $(X_1'X_1)(X_1'X_1 + X_3'X_3)^{-1}$ (see

Exercise 3.1). The following tests are often applied in practice:

(i) Roy's critical region: $Ch_1[(X_1'X_1)(X_3'X_3)^{-1}] > c$,
(ii) Lawley–Hotelling's critical region: $\mathrm{tr}(X_1'X_1)(X_3'X_3)^{-1} > c$,
(iii) Pillai's critical region: $\mathrm{tr}(X_1'X_1)(X_1'X_1 + X_3'X_3)^{-1} > c$,
(iv) The LRT with critical region: $|I + X_1(X_3'X_3)^{-1}X_1'| > c$,

where $Ch_1(\cdot)$ denotes the largest latent root of $(\cdot)$. Of course, these tests are invariant under the transformation (3.12). It is shown in Schwartz (1967a) that Pillai's test is LBI (locally best invariant). A special case of the MANOVA problem with $n_1 = 1$ and $n_2 = 0$ is known as the Hotelling T^2-problem, which is clearly invariant under the group $\mathcal{O}(n_3) \times \mathrm{GL}(p)$ acting on X by

$$g(X) = \begin{pmatrix} X_1C' \\ Q_3X_3C' \end{pmatrix}$$

and a maximal invariant is $T \equiv X_1(X_3'X_3)^{-1}X_1'$: 1×1. In this problem, Hotelling's T^2-test which rejects H for large values of T is UMPI (see Lehmann, 1959).

(A3) ANOVA

The ANOVA problem is the MANOVA problem with $p = 1$, in which $I \otimes \Sigma = \sigma^2 I_n$ with $\Sigma = \sigma^2 > 0$. Hence it is invariant under the group $\mathcal{G}$ in (3.11) with $p = 1$ and $\mathrm{GL}(p) = \{a \in R \mid a \neq 0\}$. It is well known that the LRT or the F-test based on $T = n_3X_1'X_1/n_1X_3'X_3$, which is a maximal invariant, is UMPI (see Lehmann, 1959). Obviously the Student's t-problem is a special case of the ANOVA problem.

B. Testing for Covariance Structure in (3.5) with $\Phi = I$

Most multivariate testing problems on covariance structure often treated in practice are stated as follows: Assume the MANOVA model in (3.10) so that $\Phi = I_n$ in (3.5) and test a hypothesis on Σ. Here assuming $\Phi = I_n$ in (3.5) implies the independence of the rows of X under normality. The mean structure in(3.10) coupled with invariance implies that an invariant test for Σ is based on the statistic $X_3'X_3$, whatever the hypothesis on Σ. In fact, the problem is left invariant at least under the group $\mathcal{G} = \mathcal{O}(n_3) \times R^{(n_1+n_2)\times p}$ acting on X by

$$g(X) = \begin{pmatrix} \begin{pmatrix} X_1 \\ X_2 \end{pmatrix} + F \\ Q_3X_3 \end{pmatrix} \quad \text{for} \quad g = (Q_3, F) \in \mathcal{G},$$

and a maximal invariant is $X_3'X_3$. Hence below we state some typical problems in terms of $Z \equiv X_3$ only.

(B1) Testing Independence.

The problem of testing independence between two sets of variables is to test

$$\text{H}: \Sigma_{12} = 0 \quad \text{versus} \quad \text{K}: \Sigma_{12} \neq 0 \tag{3.13}$$

in the model

$$Z = \overset{p_1\ \ p_2}{(Z_1, Z_2)} \sim N_{mp}(0, I \otimes \Sigma) \quad \text{with} \quad \Sigma = \begin{pmatrix} \Sigma_{11} & \Sigma_{12} \\ \Sigma_{21} & \Sigma_{22} \end{pmatrix} \begin{matrix} p_1 \\ p_2 \end{matrix} \tag{3.14}$$

(column blocks of Σ: p_1, p_2)

where $m \equiv n_3$. This problem is left invariant under the group $\mathscr{G} = \mathscr{O}(m) \times \text{GL}(p_1) \times \text{GL}(p_2)$ acting on Z by

$$g(Z) = (QZ_1C_1', QZ_2C_2') \quad \text{for} \quad g = (Q, C_1, C_2) \in \mathscr{G}. \tag{3.15}$$

It can be shown that a maximal invariant is the set of the ordered nonzero latent roots of $W \equiv S_{11}^{-1}S_{12}S_{22}^{-1}S_{21}$ where $S_{ij} = Z_i'Z_j$ $(i, j = 1, 2)$ (Exercise 3.2). The following invariant tests are often applied in this problem.

(i) Roy's critical region: $Ch_1(W) > c$,

(ii) Lawley–Hotelling's critical region:

$$\operatorname{tr} S_{11}^{-1}S_{12}(S_{22} + S_{21}S_{11}^{-1}S_{12})^{-1}S_{21} > c,$$

(iii) Pillai's critical region: $\operatorname{tr} W > c$,

(iv) The LRT with critical region: $|I - W| < c$.

Schwartz (1967a) showed that Pillai's test is LBI. Problem (3.13) in model (3.14) with $p_1 = 1$ is known as the R^2-problem, the problem of testing whether the multiple correlation $(\Sigma_{11}^{-1}\Sigma_{12}\Sigma_{22}^{-1}\Sigma_{21})^{1/2}$ is zero, and the test with critical region $W > c$ is UMPI. In the case of $p_1 = p_2 = 1$, the problem is to test if the ordinary bivariate correlation coefficient is zero. These problems will be treated in a nonnormal model in Section 6.1 of Chapter 6.

(B2) Testing Sphericity.

Another problem concerning covariance structure often considered in applications is testing the sphericity of Σ, i.e.,

$$\text{H}: \Sigma = \sigma^2 I \quad \text{versus} \quad \text{K}: \Sigma \neq \sigma^2 I, \sigma^2 > 0 \text{ unknown}. \tag{3.16}$$

This problem remains invariant under the group $\mathcal{G} = \mathcal{O}(m) \times \mathcal{O}(p) \times R_+$ acting on Z by $g(Z) = aPZQ'$ where $g = (P, Q, a) \in \mathcal{G}$, and a maximal invariant is the set of $r_j/\Sigma_{i=1}^{p} r_i$ $(j = 1, \ldots, p)$, where $r_1 \geq \cdots \geq r_p$ are the ordered latent roots of $S \equiv Z'Z$ (Exercise 3.3). The following invariant tests

(i) the LRT with critical region: $|S|^{m/2}/(\operatorname{tr} S)^{mp/2} < c$,
(ii) the modified LRT with critical region: $|S|^{(m-1)/2}/(\operatorname{tr} S)^{(m-1)p/2} < c$,
(iii) the LBI test with critical region: $\operatorname{tr} S^2/(\operatorname{tr} S)^2 > c$

are often used. The modified LRT is unbiased, whereas the LRT is not (see, e.g., Giri, 1977). The LBI test was derived by Sugiura (1972). This problem will be treated in context in Section 3.4 of this chapter and also in Section 6.2 of Chapter 6.

(B3) Testing $\Sigma = \Sigma_0$.

A related problem is to test

$$\text{H}: \Sigma = \Sigma_0 \qquad \text{versus} \qquad \text{K}: \Sigma \neq \Sigma_0, \tag{3.17}$$

where Σ_0 is a known p.d. matrix. Assuming $\Sigma_0 = I$ without loss of generality, the above problem is left invariant under the group $\mathcal{G} = \mathcal{O}(m) \times \mathcal{O}(p)$ acting on Z by $g(Z) = PZQ'$, where $g = (P, Q) \in \mathcal{G}$, and a maximal invariant is the set of the ordered latent roots r_i of $S = Z'Z$ (Exercise 3.4). Analogous to the problem (B2), here also the LRT with critical region $|S|^{m/2} \exp(-\frac{1}{2} \operatorname{tr} S) < c$ is biased, whereas the modified LRT with critical region $|S|^{(m-1)/2} \exp(-\frac{1}{2} \operatorname{tr} S) < c$ is unbiased (Sugiura and Nagao, 1968). For a somewhat restrictive alternative, namely, for the problem of testing

$$\text{H}: \Sigma = I \qquad \text{versus} \qquad \text{K}: \Sigma > I, \tag{3.18}$$

John (1971) proved that the test with the critical region $\operatorname{tr} S > c$ is LBI.

This problem is further discussed in Section 6.4 of Chapter 6. Some other problems on Σ in the model (3.5) with $\Phi = I$ will be considered here and there in this book.

C. *Testing for Structure of Φ in (3.5) with $p = 1$*

The problem of testing for the covariance structure of Φ in (3.5) with $p = 1$ may be regarded as a canonical form of the problem of testing for the covariance structure of error term u in a univariate linear regression model $y = R\beta + \mu$, where y: $n \times 1$. For example, the problem of testing $\rho = 0$ in the model that the error u_i's obey the AR(1) (autoregressive model of order 1)

$$u_i = \rho u_{i-1} + \varepsilon_i, \qquad |\rho| < 1 \tag{3.19}$$

is a well-known problem originally formulated by Durbin and Watson (1950, 1951). When u_i's follow the AR(1) process in (3.19), the covariance matrix of u is given by

$$\sigma^2\Phi = \frac{\sigma^2}{1-\rho^2}\begin{bmatrix} 1 & \rho & \cdots & \rho^{n-1} \\ \rho & \ddots & \ddots & \vdots \\ \vdots & \ddots & \ddots & \rho \\ \rho^{n-1} & \cdots & \rho & 1 \end{bmatrix} = \sigma^2(I - \rho A + \rho^2 B)^{-1}, \tag{3.20}$$

where $\Sigma \equiv \sigma^2$ in (3.5), $A = (a_{ij})$ is the $n \times n$ matrix with $a_{k,k-1} = a_{l,l+1} = 1$ $(k = 2,\ldots,n;\ l = 1,\ldots,n-1)$, and $B = (b_{ij})$ is the $n \times n$ matrix with $b_{ij} = 0$, except $b_{kk} = 1$ $(k = 2,\ldots,n-1)$. Other examples are the problems of testing for heteroscedasticity and intraclass covariance structure. These problems are treated in Section 3.4 of this chapter and Section 6.4 of Chapter 6.

D. *Detection of Outliers*

Tests for detection of outliers are concerned with either structure of μ (mean slippage) or structure of Φ (dispersion slippage) in the setup of (3.5). Typically, it is assumed that the mean matrix μ has a representation of the form (Ferguson, 1961, Schwager and Margolin, 1982)

$$\mu = \underset{\sim}{1}\alpha' + \Delta A \Sigma^{1/2} \tag{3.21}$$

for some unknown $\alpha \in R^p$, an arbitrary matrix $A\colon n \times p$, and the scalar parameter Δ, where $\underset{\sim}{1} = (1,\ldots,1)' \in R^n$. Under the assumption that $\Phi = I_n$, the problem is to test H: $\Delta = 0$ versus K: $\Delta \neq 0$. The matrix A, whose nonnull rows correspond to outliers when $\Delta \neq 0$, is assumed known except for the permutation of its rows.

The second version of the outlier problem is based on a structure of Φ, namely,

$$\Phi = \operatorname{diag}\{\exp(\Delta a_{v_1}),\ldots,\exp(\Delta a_{v_n})\} \equiv D_\Delta, \tag{3.22}$$

where the a's are known constants, $v = (v_1,\ldots,v_n)$ is an unknown permutation of $(1,\ldots,n)$, and $\Delta \geq 0$ is the unknown parameter controlling the presence or absence of outliers. In this formulation, since outliers tend to possess more variation than the regular observations, Δ is taken to be nonnegative and the a's are also nonnegative. The mean matrix μ is assumed to be homogeneous with respect to the different rows of X, i.e., $\mu = \underset{\sim}{1}\alpha'$ for some $\alpha \in R^p$. The problem here is to test H: $\Delta = 0$ versus K: $\Delta > 0$. If K is accepted, observations with positive a's are the outliers. However, since in our setup the permutation v is assumed unknown, we can only detect the presence or absence of outliers and not the outliers themselves.

In Chapter 7, for both problems, appropriate groups leaving the respective problems invariant are identified and the problems are analyzed in a wider framework. It turns out that the test based on Mardia's (1970) multivariate kurtosis statistic $b_{2,p}(X)$ is LBI in both cases where

(3.23)

$$b_{2,p}(X) = \sum_1^n \{(X_i - \bar{X})'S^{-1}(X_i - \bar{X})\}^2 \quad \text{and} \quad S = \sum_1^n (X_i - \bar{X})(X_i - \bar{X})',$$

where $X = (X_1, \ldots, X_n)'$, $\bar{X} = X'\underset{\sim}{1}/n$, $\underset{\sim}{1} = (1, \ldots, 1)' \in R^n$.

For $p = 1$, Ferguson (1961) proved that the test based on the sample kurtosis $b_2 = \Sigma_1^n (x_i - \bar{x})^4 / \{\Sigma_1^n (x_i - \bar{x})^2\}^2$ is LBI for detecting outliers following either (3.21) or (3.22). Schwager and Margolin (1982) extended Ferguson's result to the multivariate mean slippage case (3.21) by proving that the test based on $b_{2,p}$ is LBI.

3.3. General Theory on Null Robustness

As mentioned in Section 3.2, many multivariate parametric testing problems are usually treated in the following framework. Observe a data matrix $X: n \times p$ being distributed as an np-dimensional normal distribution $N_{np}(\mu, \Omega)$, test a certain hypothesis H on (μ, Ω), and obtain a test statistic $t(X)$. In this section, we consider the null robustness of this test statistic $t(X)$. The following two approaches are adopted for this purpose:

(1) Pick a large class of distributions containing $N_{np}(\mu, \Omega)$ under H, and derive a condition for which the distribution $\mathscr{L}(t(X))$ of $t(X)$ under H remains the same in the class, where $\mathscr{L}(\cdot)$ denotes the distribution of a random variable or matrix. (Kariya, 1981a).

(2) Fix $P_0 = N(\mu_0, \Omega_0)$, where (μ_0, Ω_0) belongs to the null hypothesis H, and obtain a condition on P for which P belongs to the class

$$\mathscr{P}_0 = \{P \in \mathscr{P} \mid \mathscr{L}(t(X) \mid P) = \mathscr{L}(t(X) \mid P_0)\},$$

where $\mathscr{P}$ is the class of probability measures on $R^{n \times p}$ and $\mathscr{L}(t(X) \mid P)$ denotes the distribution of $t(X)$ under $\mathscr{L}(X) = P \in \mathscr{P}$. (Eaton and Kariya, 1984).

***Approach* (1).** As defined in Chapter 2, an $n \times p$ random matrix X is called left-$\mathcal{O}(n)$-invariant about μ if $\mathscr{L}\{Q(X - \mu)\} = \mathscr{L}(X - \mu)$ for all $Q \in \mathcal{O}(n)$. Also, we shall call X elliptically symmetric about μ with scale matrix

$\Sigma \in \mathscr{S}(p)$ if $\mathscr{L}(gy) = \mathscr{L}(y)$ for all $g \in \mathscr{O}(np)$, where $y = (y_1, \ldots, y_n)'$, y_i is the ith row of $Y = (X - \mu)\Sigma^{-1/2}$. Let $\mathscr{X} = \{X: n \times p \,|\, \text{rank}(X) = p\}$ and assume $n \geq p$. Further, let $\mathscr{P}_{\text{L}}(\mu)$ and $\mathscr{P}_{\text{E}}(\mu, I_n \otimes \Sigma)$ denote respectively the class of np-dimensional left-$\mathscr{O}(n)$-invariant distributions about μ, such that $P(X - \mu \in \mathscr{X}) = 1$, and the class of np-dimensional elliptically symmetric distributions about μ with scale matrix $\Sigma \in \mathscr{S}(p)$, such that $P(X - \mu \in \mathscr{X}) = 1$. Clearly,

$$\mathscr{P}_{\text{E}}(\mu, I_n \otimes \Sigma) \subseteq \mathscr{P}_{\text{L}}(\mu) \qquad \text{for all} \quad \mu: n \times p \quad \text{and} \quad \Sigma \in \mathscr{S}(p).$$

If $\mathscr{L}(X) \in \mathscr{P}_{\text{E}}(\mu, I_n \otimes \Sigma)$ has a density, it is expressed as

$$f(X \,|\, \mu, \Sigma) = |\Sigma|^{-n/2} \phi(\text{tr}\, \Sigma^{-1}(X - \mu)'(X - \mu)), \tag{3.24}$$

where $\phi: [0, \infty) \to [0, \infty)$, $\int_{R^{np}} \phi(\text{tr}\, X'X)\, dX = 1$. On the other hand, if $\mathscr{L}(X) \in \mathscr{P}_{\text{L}}(\mu)$ has a density, it is expressed as

$$f(X \,|\, \mu) = \psi((X - \mu)'(X - \mu)), \tag{3.25}$$

where $\psi: \mathscr{S}(p) \to [0, \infty)$, $\int_{R^{np}} \psi(X'X)\, dX = 1$. A left-$\mathscr{O}(n)$-invariant distribution that is not elliptically symmetric is the matrix variate t-distribution, whose density is given by

$$f_0(X) = c|\Sigma|^{-n/2} |I_p + \Sigma^{-1}(X - \mu)'(X - \mu)|^{-(p+n)/2}, \tag{3.26}$$

where c is a normalizing constant. Fraser and Ng (1980) treated a multivariate regression model with this distribution for the error term.

Now in order to obtain a condition for null robustness, first assume that $\mathscr{L}(X)$ is normal with $\Omega = I \otimes \Sigma$, i.e.,

$$\mathscr{L}(X) = N(\mu, I \otimes \Sigma) \qquad \mu: n \times p, \qquad \Sigma \in \mathscr{S}(p). \tag{3.27}$$

After a suitable transformation, many null hypotheses can be stated as

$$\text{H}: (\mu, \Sigma) \in \Theta_0 \times \Lambda_0, \tag{3.28}$$

where $\Theta_0 \times \Lambda_0 \subset R^{np} \times \mathscr{S}(p)$. Here we assume

$$0 \in \Theta_0, \qquad \text{or} \qquad \Theta_0 \;\; \text{contains} \;\; 0 \;\; \text{in} \;\; R^{np}. \tag{3.29}$$

Usually this assumption is satisfied; if necessary, replace X by $X - \mu_0$ and Θ_0 by $\Theta_0 - \mu_0$, where μ_0 is a fixed point in Θ_0. Typical problems of the form (3.28) with (3.29) are the MANOVA, GMANOVA problems, the problems of testing independence or equality of covariance matrices or sphericity, etc., (see Section 3.2). In these problems, except for some special cases, there exists no UMP (uniformly most powerful) test, and consequently, many tests are proposed in each problem. A feature that these tests have in common is similarity, which is often implied by invariance. In fact, in invariant problems, the groups leaving the problems invariant often act transitively on the parameter space $\Theta_0 \times \Lambda_0$ under the null hypotheses, so that the null distributions of these tests do not depend on $(\mu, \Sigma) \in \Theta_0 \times \Lambda_0$, and the tests become similar. Now let $t(X)$ denote any statistic or vector statistic that may be a test statistic for problem (3.28) with (3.29), and consider the uniqueness of the null distributions of $t(X)$ in $\mathscr{P}_{\mathrm{L}}$ and $\mathscr{P}_{\mathrm{E}}$, where

$$\mathscr{P}_{\mathrm{L}} = \bigcup\{\mathscr{P}_{\mathrm{L}}(\mu) \mid \mu \in \Theta_0\} \tag{3.30}$$

and

$$\mathscr{P}_{\mathrm{E}} = \bigcup\{\mathscr{P}_{\mathrm{E}}(\mu, I_n \otimes \Sigma) \mid (\mu, \Sigma) \in \Theta_0 \times \Lambda_0\}. \tag{3.31}$$

Let $\mathscr{Z} = \{Z \in \mathscr{X} \mid Z'Z = I_p\}$. Let $\mathrm{GU}(p)$ (or $\mathrm{GT}(p)$) denote the set of $p \times p$ nonsingular upper (or lower) triangular matrices with positive diagonal elements. It is emphasized that in the results below, $t(X)$ can be a vector.

Theorem 3.1. *A necessary and sufficient condition for $\mathscr{L}(t(X))$ to remain the same for all $\mathscr{L}(X) \in \mathscr{P}_{\mathrm{L}}$ is that when $\mathscr{L}(X) = N(\mu, I_n \otimes \Sigma)$, the following conditions* (i) *and* (ii) *hold*:

(i) *$\mathscr{L}\{t(X - \mu)\} = \mathscr{L}\{t(X)\}$ for all $\mu \in \Theta_0$ and all $\Sigma \in \mathscr{S}(p)$.*

(ii) *$\mathscr{L}_Z\{t(ZA)\} = \mathscr{L}\{t(Z)\}$ for $\mu = 0$ and all A, $\Sigma \in \mathscr{S}(p)$, where $X = ZA$ with $Z \in \mathscr{Z}$ and $A \in \mathscr{S}(p)$, and $\mathscr{L}_Z(\cdot)$ denotes the distribution of $\cdot$ with respect to Z.*

Proof. (Sufficiency). Assume (i) and (ii) under $\mathscr{L}(X) = \mathrm{N}(\mu, I_n \otimes \Sigma)$, where $\mu \in \Theta_0$. Let $Y = X - \mu$. Since $\mathscr{L}(Y) = N(0, I_n \otimes \Sigma)$, from (i) and (ii), $\mathscr{L}\{t(X)\} = \mathscr{L}\{t(Y)\} = \mathscr{L}\{t(Z)\}$, where $Y = ZA$ with $Z \in \mathscr{Z}$ and $A \in \mathscr{S}(p)$. But by Theorem 1.1 in Chapter 1, $\mathscr{L}(Z)$ does not depend on $\mathscr{L}(Y)$ provided Y is left-$\mathscr{O}(n)$-invariant or $\mathscr{L}(Y) \in \mathscr{P}_{\mathrm{L}}(0)$. Since $\mathscr{L}(Y) \in \mathscr{P}_{\mathrm{L}}(0)$ is equivalent to $\mathscr{L}(X) \in \mathscr{P}_{\mathrm{L}}(\mu)$ with $\mu \in \Theta_0$, this implies $\mathscr{L}\{t(X)\} = \mathscr{L}\{t(Z)\}$ for all $\mathscr{L}(X) \in \mathscr{P}_{\mathrm{L}}$.

(Necessity). Since $0 \in \Theta_0$ by (3.29), $\mathscr{P}_{\mathrm{L}}(0) \subseteq \mathscr{P}_{\mathrm{L}}$. Since $\mathscr{L}(X - \mu) \in \mathscr{P}_{\mathrm{L}}(0)$ for all $\mu \in \theta_0$, the uniqueness of $\mathscr{L}\{t(X)\}$ in $\mathscr{P}_{\mathrm{L}}$ implies (i). On the other hand, since $\mathscr{L}(X) \in \mathscr{P}_{\mathrm{L}}(0)$ implies $\mathscr{L}(XC) \in \mathscr{P}_{\mathrm{L}}(0)$ for any $C \in \mathrm{GL}(p)$, the uniqueness also implies $\mathscr{L}\{t(XC)\} = \mathscr{L}\{t(X)\}$ for all $C \in \mathrm{GL}(p)$ and $\mathscr{L}(X) \in \mathscr{P}_{\mathrm{L}}$. Since $\mathscr{N} = \{N(0, I_n \otimes \Sigma) \mid \Sigma \in \mathscr{S}(p)\} \subset \mathscr{P}_{\mathrm{L}}$, in particular

$$\mathscr{L}\{t(ZAC)\} = \mathscr{L}\{t(ZA)\} \quad \text{for all} \quad C \in \mathrm{GL}(p) \quad \text{and} \quad \mathscr{L}(X) \in \mathscr{N}, \tag{3.32}$$

where by Theorem 1.1 in Chapter 1, $X = ZA$ with $Z \in \mathscr{Z}$ and $A \in \mathscr{S}(p)$, and $\mathscr{L}(Z)$ is the uniform distribution over $\mathscr{Z}$. Since $A^2 = A'Z'ZA = X'X$ and

$\mathscr{L}(X'X) = W_p(\Sigma, n)$ under $\mathscr{L}(X) = N(0, I_n \otimes \Sigma)$, A^2 is a complete sufficient statistic for $\{W_p(\Sigma, n) \mid \Sigma \in \mathscr{S}(p)\}$ where $W_p(\Sigma, n)$ denotes the Wishart distribution with mean $n\Sigma$ and degrees of freedom n. Further, since A is the unique root of A^2, A is a complete sufficient statistic. Therefore, together with the independence of Z and A and the completeness of A, (3.32) implies that for any Borel set $B \subset R^1$ and for any $C \in \mathrm{GL}(p)$

$$E^Z[I_B\{t(ZAC)\}] = E^Z[I_B\{t(ZA)\}] \qquad \text{a.e. } (A), \tag{3.33}$$

where $P\{t(ZA) \in B\} = E_\Sigma^A E^Z[I_B\{t(ZA)\}]$ is used. Here $I_B(\cdot)$ is the indicator function of B, and E^Z and E_Σ^A denote the expectations under $\mathscr{L}(Z)$ and $\mathscr{L}(A)$, respectively. Taking $C = A^{-1}$ in (3.33) yields

$$\mathscr{L}(t(Z)) = \mathscr{L}_Z(t(ZA)) \tag{3.33a}$$

a.e. (A). Since the completeness of $\mathscr{N}$ implies that of $\mathscr{P}_{\mathrm{L}}$, and since $\mathscr{L}(Z)$ remains the same for all $\mathscr{L}(X) \in \mathscr{P}_{\mathrm{L}}$ by Theorem 1.1 of Chapter 1, (3.33a) holds a.e. (A) for all $\mathscr{L}(X) = \mathscr{L}(ZA) \in \mathscr{P}_{\mathrm{L}}$. But for any given $A \in \mathscr{S}(p)$, there exists a distribution in $\mathscr{P}_{\mathrm{L}}$ which gives a positive mass to $\{A\}$. Hence (3.33a) must hold for all $A \in \mathscr{S}(p)$. This completes the proof.

We remark that in the conditions (i) and (ii) of Theorem 3.1, the part "all $\Sigma \in \mathscr{S}(p)$" can be replaced by "all $\Sigma \in \mathscr{S}$" where $\mathscr{S}$ is a nonempty open subset of $\mathscr{S}(p)$. In fact, in the proof of the necessity, even if $\mathscr{S}(p)$ is replaced by a nonempty open subset of $\mathscr{S}(p)$, the argument holds as it stands.

Corollary 3.1. *The null distribution of $t(X)$ is unique in $\mathscr{P}_{\mathrm{L}}$ if the following conditions* (i)′ *and* (ii)′ *hold*:

(i)′ $t(X - \mu) = t(X)$ *for all* $\mu \in \Theta_0$
(ii)′ $t(XC) = t(X)$ *for all* $C \in \mathscr{S}(p)$, *or for all* $C \in \mathrm{GU}(P)$ *or for all* $C \in \mathrm{GT}(p)$.

Proof. Clearly (i)′ implies (i), and from Theorem 1.1 in Chapter 1, (ii) is implied by the fact that $t(XC) = t(X)$ for all $C \in \mathscr{S}(p)$. On the other hand, by the Gram–Schmidt orthogonalization, $X = ZT$ where $Z \in \mathscr{Z}$ and $T \in \mathrm{GU}(p)$ (or $T \in \mathrm{GT}(p)$ when the orthogonalization process starts from the last column of X). Further, when $\mathscr{L}(X) = N(0, I_n \otimes \Sigma)$, $\mathscr{L}(Z)$ is the unique uniform distribution over Z; see Theorem 1.1 in Chapter 1. Hence, taking $C = T^{-1}$ in $t(ZTC) = t(ZT)$ yields $t(Z) = t(X)$, implying (ii). This completes the proof.

Corollary 3.2. *If* (i)′ *and* (ii)′ *hold*, $\mathscr{L}\{t(X)\}$ *is unique in the class* $\mathscr{P}_{\mathrm{E}}$.

Proof. This is obvious from $\mathscr{P}_E \subset \mathscr{P}_L$.

To apply Corollary 3.1 to a specific test, conditions (i)′ and (ii)′ need to be verified. Usually (i)′ is satisfied, but (ii)′ is not in some problems. In invariant problems, if the groups leaving the problems invariant contain as a subgroup $\mathscr{S}(p)$ or $\mathrm{GU}(p)$ or $\mathrm{GT}(p)$ acting on X by $X \to XC$, then the condition (ii)′ is clearly satisfied. The MANOVA and GMANOVA problems are typical examples which satisfy (ii)′.

Next, we consider the uniqueness of the null distribution of a (vector) statistic $t(X)$ in $\mathscr{P}_E$. Let $\mathscr{U} = \{U \in \mathscr{X} \mid \mathrm{tr}\, U'U = 1\}$ and $\|X\| = (\mathrm{tr}\, X'X)^{1/2}$.

Theorem 3.2. *A necessary and sufficient condition for $\mathscr{L}(t(X))$ to remain the same for all $\mathscr{L}(X) \in \mathscr{P}_E$ is that when $\mathscr{L}(X) = \mathrm{N}(\mu, I_n \otimes \Sigma)$, the following conditions* (iii) *and* (iv) *hold:*

(iii) $\mathscr{L}\{t((X-\mu)\Sigma^{-1/2})\} = \mathscr{L}\{t(X)\}$ for all $(\mu, \Sigma) \in \Theta_0 \times \Lambda_0$.
(iv) $\mathscr{L}\{t(aX/\|X\|)\} = \mathscr{L}\{t(X/\|X\|)\}$ for $a > 0$, $\mu = 0$, $\Sigma = \sigma^2 I$, $\sigma^2 > 0$.

Proof. (Sufficiency). Let $Y = (X-\mu)\Sigma^{-1/2}$ where $(\mu, \Sigma) \in \Theta_0 \times \Lambda_0$. Then from (iii) and (iv), $\mathscr{L}\{t(X)\} = \mathscr{L}\{t(Y)\} = \mathscr{L}\{t(Y/\|Y\|)\}$ since $\mathscr{L}(Y) = \mathrm{N}(0, I_n \otimes I_p)$. But $\mathscr{L}(Y/\|Y\|)$ is the unique uniform distribution over $\mathscr{U}$, provided $\mathscr{L}(Y) \in \mathscr{P}_E(0, I_n \otimes I_p)$ or equivalently $\mathscr{L}(X) \in \mathscr{P}_E(\mu, I_n \otimes \Sigma)$ with $(\mu, \Sigma) \in \Theta_0 \times \Lambda_0$ (see Chapter 1). Hence, $\mathscr{L}\{t(X)\}$ remains the same for all $\mathscr{L}(X) \in \mathscr{P}_E$.

(Necessity). Completely analogous to the proof of the necessity part of Theorem 3.1.

Corollary 3.3. *The* (*null*) *distribution of $t(X)$ is unique in $\mathscr{P}_E$ if the following conditions* (iii)′ *and* (iv)′ *hold*:

(iii)′ $t((X-\mu)\Sigma^{-1/2}) = t(X)$ for all $(\mu, \Sigma) \in \Theta_0 \times \Lambda_0$.
(iv)′ $t(\alpha X) = t(X)$ for all $\alpha > 0$.

Proof. Similar to the proof of Corollary 3.1.

We remark that the above results are extendable to the nonnull case, provided the assumption and conditions are correspondingly modified. For example, if an alternative hypothesis is of the form (3.28) and it satisfies (3.29), the above results hold exactly (see Section 3.4).

In the proof of Theorem 3.2, the result is used that the distribution of $Z/\|Z\|$ with $Z \sim N_n(0, \sigma^2 I)$ remains the same in the class of spherical distributions on

R^n which gives no mass to zero, i.e.,

$$\mathscr{S}\mathscr{P}^n = \{P \in \mathscr{P}^n \mid gP = P \quad \text{for all } g \in \mathscr{O}(n),\ P(\{0\}) = 0\},$$

where $\mathscr{P}^n$ denotes the set of all probability measures on R^n and $gP(A) = P(g^{-1}(A))$ (see Section 1.3 of Chapter 1). The result in Corollary 3.3 is nothing but a sufficient condition for a statistic to be a function of $Z/\|Z\|$ for some spherical normal variable Z.

A quick example of an application of Corollary 3.3 is the Student's t-problem in Example 3.1. There $p = 1$, $\Theta_0 = \{0\}$ in R, and the t-statistic $T(x) = \sqrt{n}\bar{x}/s$ satisfies (iii)′ and (iv)′. Hence the distribution of $T(x)$ remains the same as long as $\mathscr{L}(x) \in \mathscr{P}_{\mathrm{E}}$ with $p = 1$.

Application 3.1. *MANOVA* (see A2 in Section 3.2). A canonical form of the MANOVA model is given by (see (3.10))

$$X = (X'_1, X'_2, X'_3)' \sim N((\mu'_1, \mu'_2, 0)', \qquad I_n \otimes \Sigma)$$

where $X_i\colon n_i \times p$ $(i = 1, 2, 3)$ and $n_1 + n_2 + n_3 = n$. Here the problem is to test H: $\mu_1 = 0$ versus K: $\mu_1 \neq 0$. As has been seen in Section 3.2, the LRT, Roy's test, Lawley–Hotelling's test, and Pillai's test are functions of $t(X) = X_1(X'_3X_3)^{-1}X'_1$ where $n_3 \geq p$ is assumed. In fact, any invariant test is a function of $t(X)$, because under the group action (3.11) and (3.12), a maximal invariant is itself a function of $t(X)$. Hence showing the null robustness of the statistic $t(X)$ implies showing the null robustness of any invariant test. To show the null robustness of $t(X)$, it suffices to show that $t(X)$ satisfies conditions (i)′ and (ii)′ in Corollary 3.1. Let

$$\Theta_0 = \{(0, \mu'_2, 0)' \mid \mu_2\colon n_2 \times p\} \qquad \text{and} \quad \Lambda_0 = \mathscr{S}(p).$$

Then the null hypothesis H is of the form $(\mu, \Sigma) \in \Theta_0 \times \Lambda_0$ as in (3.28) and $0 \in \Theta_0$ satisfying (3.29). The conditions (i)′ and (ii)′ are satisfied since $t(X)$ does not depend on X_2 and since it is invariant under $X \to XC'$ with $C \in \mathrm{GL}(p)$. Therefore, we obtain

Corollary 3.4. *The null distribution of* $t(X) = X_1(X'_3X_3)^{-1}X'_1$ *remains the same as long as* $\mathscr{L}(X) \in \mathscr{P}_{\mathrm{L}}$.

Based on Dempster (1969), Dawid (1977) showed this directly.

It is noted that $t(X)$ is a maximal invariant under $\mathrm{GL}(p)$ acting on X by $X \to XC'$. This result implies that any invariant test is null-robust in the class of left-$\mathscr{O}(n)$-invariant distributions $\mathscr{P}_{\mathrm{L}}$. A member of $\mathscr{P}_{\mathrm{L}}$ is a matrix variate t-distribution with density (3.26).

The null robustness of invariant tests in the GMANOVA problem is considered in Chapter 5.

Application 3.2. *Tests of Independence* (see B1 in Section 3.2). In the reduced model $Z \sim N(0, I_m \otimes \Sigma)$ as described in (3.14), the problem of testing independence deals with $\mathrm{H}: \Sigma_{12} = 0$, where $\Sigma = (\Sigma_{ij})$ with $\Sigma_{12}: p_1 \times p_2$. Invariant tests are functions of $S_{11}^{-1} S_{12} S_{22}^{-1} S_{21}$, where $S_{ij} = Z_i' Z_j$ with $Z = (Z_1, Z_2)$ and $Z_i: m \times p_i$ $(i, j = 1, 2)$. Since $\mathscr{L}(S)$ does not depend on μ_i's in (3.10), condition (i) in Theorem 3.1 is satisfied. Here $S = (S_{ij})(i, j = 1, 2)$. On the other hand, condition (ii) is not satisfied unless the test is trivial. To see this, suppose a test statistic $t(X)$ based on S satisfies (ii). Then, by Theorem 3.1, $\mathscr{L}(t(X))$ remains the same for all $\Sigma \in \mathscr{S}(p)$. Since $\Sigma_{12} = 0$ versus $\Sigma_{12} \neq 0$ is tested, this implies that the power function of the test is equal to the significance level. Therefore any test based on S, including the LRT, Roy's test, Lawley–Hotelling's test, and Pillai's test, does not have unique null distributions in $\mathscr{P}_{\mathrm{L}}$. However, it is easy to check that the conditions (iii)$'$ and (iv)$'$ in Corollary 3.3 are satisfied for all invariant tests including these tests because $t(\alpha X) = t(X)$ for all $\alpha > 0$. Hence invariant tests have unique null distributions in the class of elliptically symmetric distributions $\mathscr{P}_{\mathrm{E}}$. A bigger class of the distributions for which the null robustness of an invariant test obtains is derived by Approach (2) below.

Application 3.3. *Tests for Sphericity.* As in (B2) of Section 3.2, invariant tests for the hypothesis of sphericity, namely $\mathrm{H}: \Sigma = \sigma^2 I_m$ in the reduced model Z: $m \times p \sim N(0, I_m \otimes \Sigma)$ are functions of the maximal invariant $t(Z) = \{r_1/\Sigma r_j, \ldots, r_p/\Sigma r_j\}$ where the r_j's are the latent roots of $S = Z'Z$. It is easy to see that $t(Z)$ satisfies the conditions (iii)$'$ and (iv)$'$. Hence the null robustness in $\mathscr{P}_{\mathrm{E}}$ of all the invariant tests follows.

Application 3.4. *Tests for Equality of Covariance Matrices.* Let the X_i's be independent normal random matrices:

$$X_i \sim N(e_i \mu', I_{n_i} \otimes \Sigma_i),$$

where X_i: $n_i \times p$, $e_i = (1, \ldots, 1)' \in R^{n_i}$, $\mu \in R^p$, and $\Sigma_i \in \mathscr{S}(p)$ $(i = 1, \ldots, k)$. The problem here is to test

$$\mathrm{H}: \Sigma_1 = \Sigma_2 = \cdots = \Sigma_k.$$

This null hypothesis is of the form (3.28) with (3.29), where $\Sigma = \Sigma_1$. In fact,

$$X = (X_1', \ldots, X_k')' \sim N(e\mu', I_n \otimes \Sigma)$$

under H, where $e = (e_1', \ldots, e_k')'$ and $n_1 + \cdots + n_k = n$. Clearly this problem is invariant under the group of transformations $X_i \to Q_i X_i C' + e_i a'$, where $Q_i \in \mathcal{O}(n_i)$, $C \in \mathrm{GL}(p)$ and $a \in R^p$, and the group contains as a subgroup $\mathrm{GL}(p)$ acting on X by $X \to XC'$. Therefore the conditions (i)′ and (ii)′ in Corollary 3.1 are satisfied, and so the null distributions of invariant tests are unique in the class of left-$\mathcal{O}(n)$-invariant distributions $\mathscr{P}_{\mathrm{L}}$. This fact was pointed out by Chmielewski (1980) after the uniqueness in $\mathscr{P}_{\mathrm{E}}$ was directly shown.

Application 3.5. *Tests for Outliers.* As mentioned in subsection D of Section 3.2 and to be proved in Chapter 7, the LBI test for the detection of outliers under the formulation (3.21) or (3.22) depends on the statistic $t(X) = b_{2,p}(X)$ given in Section 3.2. It is easy to verify that $t(X)$ satisfies (i)′ and (ii)′. Hence the null distribution of $b_{2,p}(X)$ remains the same as long as $\mathscr{L}(X) \in \mathscr{P}_{\mathrm{L}}$.

Many other testing problems are treated in different textbooks as well as in Kiefer and Schwartz (1965), and in Krishnaiah (1978). For each problem, by checking the conditions for the uniqueness in $\mathscr{P}_{\mathrm{E}}$ or $\mathscr{P}_{\mathrm{L}}$ as demonstrated above, it can be verified if a given test has the unique null distribution in $\mathscr{P}_{\mathrm{E}}$ or $\mathscr{P}_{\mathrm{L}}$.

We remark that, as has been observed in the problem of testing independence, if a problem concerns the structure of the covariance matrix Σ in the model $N(0, I_n \otimes \Sigma)$, the condition (ii) in Theorem 3.1 is not satisfied. In other words, unless a null hypothesis on Σ contains a nonempty open set of $\mathscr{S}(p)$, the condition (ii) is not satisfied. On the contrary, almost all similar tests that we see in applications satisfy the conditions (iii)′ and (iv)′ in Corollary 3.3. It may be pointed out that our formulation does not treat a testing problem dealing with a hypothesis of the form $(\theta, \Sigma) \in \Delta \subset R^{np} \times \mathscr{S}(p)$, such that Δ cannot be expressed as $\Delta = \Theta_0 \times \Lambda_0$ for any $\Theta_0 \subset R^{np}$ and $\Lambda_0 \subset \mathscr{S}(p)$.

Approach (2). The approach here is associated with the underlying structure of Approach (1). The following example in conjunction with Theorem 1.1 of Chapter 1, which we used in Approach (1), illustrates the basic idea.

Example 3.2. Take $\mathscr{Y}$ to be the set of all $n \times p$ matrices of rank p and consider a random matrix $Z \in \mathscr{Y}$. A number of important statistics which arise in testing problems in MANOVA model (see Section 3.2) can be written as functions of

$$T = t(Z) = Z(Z'Z)^{-1}Z',$$

which is a random projection of rank p. The distributions of these functions are often computed under $\mathscr{L}(Z) = P_0$ where $P_0 = N(0, I_n \otimes I_p)$. Here we claim that if $\mathscr{L}(\Gamma Z) = \mathscr{L}(Z)$ for each $\Gamma \in \mathcal{O}(n)$, then $\mathscr{L}(Z)$ belongs to

$$(*) \qquad \{P \mid \mathscr{L}(t(Z) \mid P) = \mathscr{L}(t(Z) \mid P_0)\},$$

where $\mathscr{L}(t(Z) \mid P)$ is the distribution of $t(Z)$ under $\mathscr{L}(Z) = P$. To see this, first observe that $t(\Gamma Z) = \Gamma t(Z)\Gamma'$, so

$$\mathscr{L}(T) = \mathscr{L}(\Gamma T \Gamma'), \qquad \Gamma \in \mathcal{O}(n),$$

since $\mathscr{L}(\Gamma Z) = \mathscr{L}(Z)$ for $\Gamma \in \mathcal{O}(n)$. However, this characterizes the distribution of T because: (i) $\mathcal{O}(n)$ is compact, and (ii) $\mathcal{O}(n)$ acts transitively on the set of $n \times n$ rank p projections (see Chapter 2). In fact, let $\Psi \in \mathcal{O}(n)$ be such that $\Psi' T \Psi = J = \text{diag}\{1, \ldots, 1, 0, \ldots, 0\}$, and take a random matrix $\Gamma \in \mathcal{O}(n)$ having the invariant probability measure over $\mathcal{O}(n)$, which is independent of T. Then from (*) and the invariance of $\mathscr{L}(\Gamma)$,

$$\mathscr{L}(T) = \mathscr{L}(\Gamma \Psi J \Psi' \Gamma') = \mathscr{L}(\Gamma J \Gamma').$$

This implies that the distribution of T is induced by the invariant probability measure of Γ. Thus $\mathscr{L}(T)$ remains the same so long as $\mathscr{L}(\Gamma Z) = \mathscr{L}(Z)$ for each $\Gamma \in \mathcal{O}(n)$.

This example contains the elements of the argument which leads to Theorem 3.3 below.

Suppose that $(\mathscr{X}, \mathscr{B})$ is a measurable space and $\mathscr{G}_0$ is a group that acts measurably on the left of $\mathscr{X}$. Let $\mathscr{P}$ be the set of all probability measures on $(\mathscr{X}, \mathscr{B})$. If $P \in \mathscr{P}$ and $g \in \mathscr{G}_0$, then gP denotes the probability defined by

$$(gP)(B) = P(g^{-1}B), \; B \in \mathscr{B}.$$

The basic situation to be considered here is the following: A measurable space $(\mathscr{X}, \mathscr{B})$ is acted on measurably and transitively by a locally compact topological group $\mathscr{G}$. It is assumed that the group $\mathscr{G}$ can be represented as $\mathscr{G} = \mathscr{K} \cdot \mathscr{H}$ (each g in $\mathscr{G}$ can be written $g = kh$ for $k \in \mathscr{K}$, $h \in \mathscr{H}$), where

(i) $\mathscr{H}$ is a normal subgroup of $\mathscr{G}$,
(ii) $\mathscr{K}$ is a compact subgroup of $\mathscr{G}$.

In most applications $\mathscr{K}$ will be the quotient group $\mathscr{G}/\mathscr{H}$, but that assumption is not necessary here. Let λ denote the unique invariant probability measure on $\mathscr{K}$. We assume that the mapping $(k, x) \to kx$ from $\mathscr{K} \times \mathscr{X}$ into $\mathscr{X}$ is a jointly measurable mapping. Also, let

$$\mathscr{P}_{\mathscr{K}} = \{P \in \mathscr{P} \mid kP = P \quad \text{for all} \quad k \in \mathscr{K}\},$$

which denotes the class of all $\mathscr{K}$-invariant distributions.

Theorem 3.3. *Suppose that t is a measurable mapping from $(\mathscr{X}, \mathscr{B})$ to $(\mathscr{Y}, \mathscr{C})$ such that t is $\mathscr{H}$-invariant. Then,*

$$\mathscr{L}(t(X) \mid P) = \mathscr{L}(t(X) \mid P') \quad \text{for} \quad P, P' \in \mathscr{P}_{\mathscr{K}}.$$

Proof. It suffices to show that for each bounded measurable real-valued function f, we have

$$(3.34) \qquad \int f(t(x))P(dx) = \int f(t(x))P'(dx) \quad \text{for} \quad P, P' \in \mathscr{P}_{\mathscr{K}}.$$

Since $P \in \mathscr{P}_{\mathscr{K}}$, for $k \in \mathscr{K}$ it follows that

$$(3.35) \qquad \int f(t(x))P(dx) = \int f(t(kx))P(dx).$$

Integrating both sides of (3.35) over $\mathscr{K}$ yields

$$(3.36) \qquad \int f(t(x))P(dx) = \int_{\mathscr{K}} \int_{\mathscr{X}} f(t(kx))P(dx)\lambda(dk).$$

Let $x_0 \in \mathscr{X}$. Given $x \in \mathscr{X}$, there exists $g \in \mathscr{G}$ such that $x = gx_0$, since $\mathscr{G}$ acts transitively on $\mathscr{X}$. By assumption, $g = k_1 h$ for some $k_1 \in \mathscr{K}$ and $h \in \mathscr{H}$. Thus,

$$(3.37) \qquad f(t(kx)) = f(t(kk_1 hx_0)) = f(t(kk_1 h(kk_1)^{-1} kk_1 x_0)) = f(t(kk_1 x_0)),$$

where the last equality follows from the normality of $\mathscr{H}$ and the assumed invariance of t. From (3.37) and the invariance of the measure λ, we have

$$(3.38) \qquad \int_{\mathscr{K}} f(t(kx))\lambda(dk) = \int_{\mathscr{K}} f(t(kk_1 x_0))\lambda(dk) = \int_{\mathscr{K}} f(t(kx_0))\lambda(dk).$$

Using Fubini's Theorem and substituting (3.38) into (3.36) yields

$$(3.39) \qquad \int_{\mathscr{X}} f(t(x))P(dx) = \int_{\mathscr{X}} \int_{\mathscr{K}} f(t(kx_0))\lambda(dk)P(dx) = \int_{\mathscr{K}} f(t(kx_0))\lambda(dk).$$

Since (3.39) holds for each $P \in \mathscr{P}_{\mathscr{K}}$, (3.34) holds and the proof is complete.

In some situations (see Application 3.7), the subgroup $\mathscr{H}$ of interest is not normal in $\mathscr{G}$, but the compact subgroup $\mathscr{K}$ is normal in $\mathscr{G}$. In this case, we still have the conclusion of Theorem 3.3. More precisely, assume that $\mathscr{G}$ acts transitively on $\mathscr{X}$ and that $\mathscr{G} = \mathscr{K} \cdot \mathscr{H}$ where

(iii) $\mathscr{K}$ is normal in $\mathscr{G}$,
(iv) $\mathscr{K}$ is a compact subgroup of $\mathscr{G}$.

Theorem 3.4. *Suppose that the measurable map t from $(\mathscr{X}, \mathscr{B})$ to $(\mathscr{Y}, \mathscr{C})$ is invariant under $\mathscr{H}$. Then*

$$\mathscr{L}(t(X) \mid P) = \mathscr{L}(t(X) \mid P') \qquad \text{for} \quad P, P' \in \mathscr{P}_{\mathscr{K}}.$$

Proof. The proof is similar to that of Theorem 3.3. It suffices to establish (3.34). The argument used to prove Theorem 3.3 shows that a given x can be written as $k_1 h x_0$ where x_0 is fixed in $\mathscr{X}$. Thus, for any bounded measurable f on $\mathscr{Y}$, we have

$$f(t(kx)) = f(t(kk_1 h x_0)) = f(t(h^{-1} k k_1 h x_0)), \tag{3.40}$$

since t is $\mathscr{H}$-invariant. Since equation (3.36) is valid for the case at hand, Fubini's Theorem and (3.40) yield

$$\begin{aligned} \int f(t(x))P(dx) &= \int_{\mathscr{K}} \int_{\mathscr{X}} f(t(kx))P(dx)\lambda(dk) \\ &= \int_{\mathscr{X}} \int_{\mathscr{K}} f(t(h^{-1}kk_1hx_0))\lambda(dk)P(dx). \end{aligned} \tag{3.41}$$

However, the normality of $\mathscr{K}$ in $\mathscr{G}$ and the invariance of the probability measure λ on $\mathscr{K}$ implies that

$$\int_{\mathscr{K}} f(t(h^{-1}kk_1hx_0))\lambda(dk) = \int_{\mathscr{K}} f(t(kx_0))\lambda(dk).$$

Combining this with (3.18) yields

$$\int_{\mathscr{X}} f(t(x))P(dx) = \int_{\mathscr{K}} f(t(kx_0))\lambda(dk).$$

Thus equation (3.34) holds, and the proof is complete.

Application 3.6. *MANOVA.* To demonstrate an application of Approach (2), let us again consider the MANOVA problem treated in Application 3.1 of this section. Recall that a maximal invariant is a function of $t(X) = X_1(X'_3X_3)^{-1}X'_1$. Here we consider the statistic

$$t_0(X) = \begin{pmatrix} t_1(X) \\ t_2(X) \end{pmatrix} = \begin{pmatrix} X_1(X'_1X_1 + X'_3X_3)^{-1}X'_1 \\ X_3(X'_1X_1 + X'_3X_3)^{-1}X'_3 \end{pmatrix} = \begin{pmatrix} X_1(W'W)^{-1}X'_1 \\ X_3(W'W)^{-1}X'_3 \end{pmatrix}$$

where $W = (X_1', X_3')'$. For convenience, write

$$X = (X_2', W')' = (X_2', (X_1', X_3'))'.$$

Noting that $t(X)$ is in one–one correspondence with $t_1(X)$, in order to show the null robustness of $t(X)$, it suffices to show the null robustness of $t_0(X)$. Let $\mathscr{X}$ be the space of X or $R^{n\times p}$ and let $P_0 = N(0, I_n \otimes I_p)$. It is easily shown that $\mathscr{L}(t_0(X)\,|\,P_0)$ is the same as $\mathscr{L}(t_0(X)\,|\,P_1)$ when $P_1 = N((\mu_2', 0, 0)', I_n \otimes \Sigma)$ for any $\mu_2 \in R^{n_2\times p}$ and $\Sigma \in \mathscr{S}(p)$. Now in order to apply Theorem 3.3, consider the group $\mathscr{G}$ whose elements are (Γ, A, F) where $\Gamma \in \mathscr{O}(n_1 + n_3)$, $A \in \mathrm{GL}(p)$ and $F \in R^{n_2\times p}$. Here it is noted that this group is slightly different from the group $\mathscr{G}$ in (3.11) of Section 2, which leaves the MANOVA problem invariant. The action of $\mathscr{G}$ on $\mathscr{X}$ is

$$(\Gamma, A, F)(X) = (X_2 A' + F, \Gamma W A'),$$

and the group operation is

$$(\Gamma_1, A_1, F_1)(\Gamma_2, A_2, F_2) = (\Gamma_1\Gamma_2, A_1 A_2, F_2 A_1' + F_1).$$

Then the action of $\mathscr{G}$ is transitive. With

$$\mathscr{H} = \{(\Gamma, A, F) \in \mathscr{G}\,|\,\Gamma = I_{n_1+n_3}\}$$

and

$$\mathscr{K} = \{(\Gamma, A, F) \in \mathscr{G}\,|\,A = I_p,\, F = 0\},$$

it follows that $\mathscr{G} = \mathscr{K}\cdot\mathscr{H}$, $\mathscr{H}$ is normal in $\mathscr{G}$, and $\mathscr{K}$ is compact.

Let $\mathscr{Y}$ be the set of all $(n_1 + n_3) \times (n_1 + n_3)$ rank p orthogonal projections on $R^{n_1+n_3}$ (and equip $\mathscr{Y}$ with the usual topology). Then the function $t_0(X)$ is a map from $\mathscr{X}$ onto $\mathscr{Y}$, is measurable, and $\mathscr{H}$-invariant. Hence by Theorem 3.3, $\mathscr{L}(t_0(X)\,|\,P) = \mathscr{L}(t_0(X)\,|\,P_0)$ for any $P \in \mathscr{P}_{\mathscr{K}}$. In particular, if the distribution of X satisfies

$$\mathscr{L}(X) = \mathscr{L}((X_2', (\Gamma W)')') \qquad \text{for any} \quad \Gamma \in \mathscr{O}(n_1 + n_3), \tag{3.42}$$

then the distribution of $t_0(X)$ (henceforth $t_1(X)$) is the same as if $\mathscr{L}(X) = N(0, I_n \otimes I_p)$. Since $t(X)$ is a bijection of $t_1(X)$, all invariant tests are null-robust in the class $\mathscr{P}_{\mathscr{K}}$. An important observation is that the class of distributions satisfying (3.42), which is included in $\mathscr{P}_{\mathscr{K}}$, includes the class of left-$\mathscr{O}(n)$-invariant distributions $\mathscr{P}_{\mathrm{L}}$.

Application 3.7. *Canonical Correlations.* Consider the null robustness of canonical correlations (see Application 3.2 in Approach (1)). Without essential loss of generality, we consider the mean zero case: $Z \sim N(0, I_m \otimes \Sigma)$ (see Section 3.2). The sample space for this section is $\mathscr{X}$, the set of $m \times p$ matrices of rank p. Partition Z as $Z = (Z_1, Z_2)$, where Z_i is $m \times p_i$, $i = 1, 2$. The orthogonal projections

$$Q_i = Z_i(Z_i'Z_i)^{-1}Z_i', \qquad i = 1, 2,$$

are elements of $\mathscr{Y}_{m,p_i}$, $i = 1, 2$, where $\mathscr{Y}_{m,p}$ is the space of $m \times m$ orthogonal projections of rank p. The squared canonical correlations are defined to be the $q \equiv \min\{p_1, p_2\}$ largest eigenvalues of Q_1Q_2. To see that this definition agrees with the standard definition in terms of the sample covariance matrix $S = Z'Z$, partition S as $S = (S_{ij})$ with S_{ij}: $p_i \times p_j$ for $i, j = 1, 2$. Classically, the squared canonical correlations are defined to be the q-largest eigenvalues of $S_{11}^{-1}S_{12}S_{22}^{-1}S_{21}$. But $S_{ij} = Z_i'Z_j$, so

$$Q_1Q_2 = Z_1S_{11}^{-1}S_{12}S_{22}^{-1}Z_2'.$$

However, the nonzero eigenvalues of $Z_1S_{11}^{-1}S_{12}S_{22}^{-1}Z_2'$ are the same as the nonzero eigenvalues of

$$S_{11}^{-1}S_{12}S_{22}^{-1}Z_2'Z_1 = S_{11}^{-1}S_{12}S_{22}^{-1}S_{21}.$$

Thus, our definition coincides with the usual definition.

Given $Z \in \mathscr{X}$, let $t(Z)$ be the vector of the q-largest eigenvalues (arranged in order) of Q_1Q_2. When $\mathscr{L}(Z)$ is $N(0, I_m \otimes I_p) \equiv P_0$, the density of $t(Z)$ is known (see Anderson, 1958, Chapter 13). Here we shall describe a large class of distributions under which $t(Z)$ has the same distribution as it does when $\mathscr{L}(Z) = P_0$. Consider the group $\mathscr{G} = \mathscr{O}(m) \dot{\times} \mathscr{O}(m) \times \mathrm{GL}(p_1) \times \mathrm{GL}(p_2)$ whose elements are (Ψ, Γ, A, B) with Ψ, $\Gamma \in \mathscr{O}(m)$, $A \in \mathrm{GL}(p_1)$, and $B \in \mathrm{GL}(p_2)$. The action of $\mathscr{G}$ on $\mathscr{X}$ is

$$(\Psi, \Gamma, A, B)(z_1, z_2) = (\Psi z_1 A', \Gamma z_2 B'),$$

and $\mathscr{G}$ acts transitively on $\mathscr{X}$. The group operation is

$$(\Psi_1, \Gamma_1, A_1, B_1)(\Psi_2, \Gamma_2, A_2, B_2) = (\Psi_1\Psi_2, \Gamma_1\Gamma_2, A_1A_2, B_1B_2).$$

Let

$$\mathscr{H} = \{(\Psi, \Gamma, A, B) \in \mathscr{G} \mid \Psi = \Gamma\}$$

and

$$\mathscr{K} = \{(\Psi, \Gamma, A, B) \in \mathscr{G} \mid \Gamma = I_m, A = I_{p_1}, B = I_{p_2}\}.$$

Then $\mathscr{G} = \mathscr{K} \cdot \mathscr{H}$, $\mathscr{K}$ is compact and $\mathscr{K}$ is normal in $\mathscr{G}$. Thus Theorem 3.4 is applicable, and we have

$$\mathscr{L}(t(X) \mid P_0) = \mathscr{L}(t(Z) \mid P),$$

for all $P \in \mathscr{P}_{\mathscr{K}}$, since $P_0 \in \mathscr{P}_{\mathscr{K}}$. To describe $\mathscr{P}_{\mathscr{K}}$, first observe that the elements of $\mathscr{K}$ act on $\mathscr{X}$ by

$$(\Psi, I_m, I_{p_1}, I_{p_2})(z_1, z_2) = (\Psi z_1, z_2).$$

Thus $\mathscr{L}(Z) \in \mathscr{P}_{\mathscr{K}}$ if and only if $Z = (Z_1, Z_2)$ has the same distribution as $(\Psi Z_1, Z_2)$, $\Psi \in \mathcal{O}(m)$. This certainly occurs when $\mathscr{L}(Z) = N(O, I_m \otimes \bar{\Sigma})$, where

$$\bar{\Sigma} = \begin{pmatrix} \Sigma_{11} & 0 \\ 0 & \Sigma_{22} \end{pmatrix}, \Sigma_{ii}: p_i \times p_i, \qquad i = 1, 2.$$

When Z has a density on $\mathscr{X}$ which can be written as

$$f(z_1, z_2) = f_1(z_1' z_1, z_2), \tag{3.43}$$

for some f_1, then it is easy to show that

$$\mathscr{L}(Z_1, Z_2) = \mathscr{L}(\Psi Z_1, Z_2), \tag{3.44}$$

for $\Psi \in \mathcal{O}(m)$. (When Z has a density, condition (3.43) is almost necessary for $\mathscr{L}(Z_1, Z_2) = \mathscr{L}(\Psi Z_1, Z_2)$ with measure-theoretic difficulties being the only problem). It should be noted that, in this example, $\mathscr{H}$ is not normal in $\mathscr{G}$, and condition (i) fails to hold for the statistic $t(Z)$. Thus, it makes no sense to speak of $\mathscr{K}$ including a transitive action on the space of t values.

3.4. General Theory on Nonnull Robustness

In this section, when a testing problem is left invariant under a group, sufficient conditions are given on the structure of a model for which the nonnull robustness of an invariant test in the sense of Definition 3.2. holds. As applications, the nonnull robustness of some invariant tests for certain covariance structures is shown. All the notions on invariance in this section are defined in Chapter 2.

Recall that the standard invariance approach to a testing problem requires first to find a group leaving the problem invariant, second, to find a maximal invariant, third, to derive the distribution of the maximal invariant, and, thereafter, to proceed to derive an optimum test based on the distribution. Hence, to establish the nonnull robustness of an invariant test, it is sufficient to show the nonnull robustness of a maximal invariant, because any invariant test is a function of the maximal invariant. Here, on the basis of the representation theorem on the probability ratio of a maximal invariant discussed in Section 2.3 of Chapter 2, we give a sufficient condition for the nonnull robustness of a maximal invariant. Consequently if the condition is satisfied, all the invariant tests are nonnull-robust.

Let $\mathscr{X}$ be a nonempty open subset of R^n with the Borel sigma-field $\mathscr{B}$ and let $\mathscr{G}$ be a locally compact and sigma-compact group acting on the left of $\mathscr{X}$. Let $\mathscr{F}(\Theta)$ be a class of densities on $\mathscr{X}$ with respect to a relatively left-invariant Borel measure μ with left multiplier χ, and assume that each density $f(\cdot \mid \theta)$ in $\mathscr{F}(\Theta)$ is of the form

$$(3.45) \qquad f(x \mid \theta) = \beta(\theta) q(\psi(x : \theta)), \qquad \theta \in \Theta$$

where, as defined in Section 2.2 of Chapter 2, μ and χ satisfy $\mu(gE) = \chi(g)\mu(E)$ for any $g \in \mathscr{G}$ and $E \in \mathscr{B}$. Here it is assumed that $\beta(\cdot)$ is a known function, $\psi(\cdot : \theta)$ is a known measurable function from $\mathscr{X}$ onto $\mathscr{Y}$, $\mathscr{Y}$ is a nonempty open subset of R^m and independent of θ, and Θ is a nonempty open subset of R^p. Further suppose that the group $\mathscr{G}$ leaves invariant the following problem

$$(3.46) \qquad \text{H}: \theta \in \Theta_0 \qquad \text{versus} \qquad \text{K}: \theta \in \Theta_1,$$

where $\Theta_0 \cap \Theta_1 = \phi$ and $\Theta = \Theta_0 \cup \Theta_1$. Here we assume that $\mathscr{X}$ is a proper $\mathscr{G}$ space (see Definition 2.9 of Chapter 2). Then by Theorem 2.4 of Chapter 2, the probability ratio R of the maximal invariant $\pi = \pi(x)$ under $\theta_1 \in \Theta_1$ and $\theta_0 \in \Theta_0$ is given by

$$(3.47) \qquad R = R(\pi(x)) \equiv (dP^{\pi}_{\theta_1}/dP^{\pi}_{\theta_0})(\pi(x)) = H(x \mid \theta_1)/H(x \mid \theta_0)$$

with

$$(3.48) \qquad H(x \mid \theta) = \int_{\mathscr{G}} f(gx \mid \theta)\chi(g)\nu(dg),$$

where $f(x \mid \theta)$ is given by (3.45) and $\nu(dg)$ is a left-invariant measure on $\mathscr{G}$. It is noted that $H(x \mid \theta_1) < \infty$ and $H(x \mid \theta_0) \neq 0$ a.e. $(P^{\pi}_{\theta_0})$, and that π is the projection map from $\mathscr{X}$ onto the quotient space $\mathscr{X}/\mathscr{G}$ with $\pi(x) = \mathscr{G}x$. Also note that any maximal invariant is a bimeasurable bijection of π. Now suppose that R in (3.47) is independent of q for all $\theta_1 \in \Theta_1$ and $\theta_0 \in \Theta_0$, and that the null distribution $P^{\pi}_{\theta_0}$ of $\pi = \pi(x)$ does not depend on q in (3.45) for each θ_0. Then the nonnull distribution $dP^{\pi}_{\theta_1} = R\,dP^{\pi}_{\theta_0}$ of π is independent of q. Hence the nonnull distribution of any invariant test statistic is independent of q, since an invariant statistic is a measurable function of π. This implies the nonnull robustness of an invariant test under $\mathscr{G}$. Therefore we shall give a sufficient condition for R in (3.47) to be independent of q. For this purpose, we make the following two assumptions.

Assumption 3.1. $\mathscr{G}$ is a product group $\mathscr{G} = \mathscr{G}_1 \times \mathscr{G}_2$ with product left-invariant measure $\nu = \nu_1 \times \nu_2$ and product left multiplier $\chi(g) = \chi_1(g_1)\chi_2(g_2)$ for $g = (g_1, g_2) \in \mathscr{G}$.

Assumption 3.2. The function ψ in (3.45) satisfies

$$\psi((g_1, g_2)x:\theta) = \bar{g}_1\psi((e_1, g_2)x:\theta) \tag{3.49}$$

for all $x \in \mathcal{X}, (g_1, g_2) \in \mathcal{G}_1 \times \mathcal{G}_2$ and $\theta \in \Theta$, and $\bar{\mathcal{G}}_1$ acts transitively on the range space $\mathcal{Y}$ of ψ, where $\bar{\mathcal{G}}_1$ is an induced group of $\mathcal{G}_1$ as a continuous homomorphic image, $\bar{g}_1 \in \bar{\mathcal{G}}_1$, and e_i is the unit element of $\mathcal{G}_i$, $i = 1, 2$.

Theorem 3.5. *Under the Assumptions* 3.1 *and* 3.2, *the probability ratio* R *in* (3.47) *is independent of* q *for all* $\theta_i \in \Theta_i$ $(i = 0, 1)$. *In fact, it is given by* $R = K(x\,|\,\theta_1)/K(x\,|\,\theta_0)$ *with*

$$K(x\,|\,\theta) = \beta(\theta)\int_{\mathcal{G}_2} \chi_1(h(x, g_2, \theta:y_0))\Delta_1(h(x, g_2, \theta:y_0))\chi_2(g_2)\nu_2(dg_2), \tag{3.50}$$

where y_0 *is a fixed element of* $\mathcal{Y}$, $h(x, g_2, \theta:y_0)$ *is an element of* $\mathcal{G}_1$ *such that*

$$\overline{h(x, g_2, \theta:y_0)}\psi((e_1, g_2)x:\theta) = y_0, \tag{3.51}$$

and Δ_1 *is a right-modular function of* ν_1. *Here the measurability of* $h(x, g_2, \theta:y_0)$ *in* g_2 *is assumed.*

Proof. From (3.45), (3.48) and (3.49), the numerator of R is given by

$$\begin{aligned} H(x\,|\,\theta_1) &= \int_{\mathcal{G}_1\times\mathcal{G}_2} \beta(\theta_1)q(\psi(gx:\theta_1))\chi(g)\nu(dg) \\ &= \int_{\mathcal{G}_2} H_1(x, g_2\,|\,\theta_1)\chi_2(g_2)\nu_2(dg_2) \end{aligned} \tag{3.52}$$

with

$$H_1(x, g_2\,|\,\theta_1) = \beta(\theta_1)\int_{\mathcal{G}_1} q[\bar{g}_1\psi((e_1, g_2)x:\theta_1)]\chi_1(g_1)\nu_1(dg_1). \tag{3.53}$$

Since $\bar{\mathcal{G}}_1$ acts transitively on the range space $\mathcal{Y}$ of $\psi(x:\theta_1)$, for any $y_0 \in \mathcal{Y}$ there exists $h(x, g_2, \theta_1:y_0) \in \mathcal{G}_1$ such that (3.51) with $\theta = \theta_1$ is satisfied. Hence using the invariance of ν_1 and replacing g_1 by $g_1 h(x, g_2, \theta_1:y_0)$ in (3.53) yields

$$H_1(x, g_2\,|\,\theta_1) = \beta(\theta_1)\chi_1(h(x, g_2, \theta_1:y_0))\Delta_1(h(x, g_2, \theta_1:y_0))k, \tag{3.54}$$

with $k = \int_{\mathcal{G}_1} q(g_1 y_0)\chi_1(g_1)\nu_1(dg_1)$. This implies $H(x\,|\,\theta_1) = kK(x\,|\,\theta_1)$ from (3.50) and (3.52). In the same way, we obtain $H(x\,|\,\theta_0) = kK(x\,|\,\theta_0)$. Therefore the ratio R in (3.47) becomes $K(x\,|\,\theta_1)/K(x\,|\,\theta_0)$, completing the proof.

It is noted that no assumption is made on the form of q in (3.45).

In the situation in which the assumptions in Theorem 3.5 hold and P_θ^π is independent of q for each $\theta \in \Theta_0$, the distribution of a maximal invariant statistic is completely independent of the form of q for each parameter θ, since any maximal invariant is a bimeasurable bijection of the maximal invariant π. Hence the whole invariance-reduced decision problem is also independent of the form of q. This implies that all decision-theoretic properties via invariance are robust: unbiasedness, Bayes, admissibility, minimaxity, etc. In particular, as was pointed out in Section 3.1, the optimality robustness follows automatically.

Application 3.8. *Tests for Covariance Structure in Regression.* Let us consider a regression model

$$y = X\beta + u, \tag{3.55}$$

where X is an $n \times k$ fixed matrix of rank k, and the error term $u: n \times 1$ is assumed to follow an elliptically contoured distribution with a density of the form

$$f(u \mid \sigma^2\Sigma) = |\sigma^2\Sigma|^{-1/2} q(u'\Sigma^{-1}u/\sigma^2). \tag{3.56}$$

Here, as discussed in detail in Chapter 1, q is a function from $[0, \infty)$ into $[0, \infty)$, such that $\int_{R^n} q(u'u)\, du = 1$, and $\Sigma \in \mathcal{S}(n)$ is the scale matrix. Under this setup, we consider the testing problem

$$\mathrm{H}_0: \Sigma \in \Lambda_0 \qquad \text{versus} \qquad \mathrm{H}_1: \Sigma \in \Lambda_1, \tag{3.57}$$

where $\Lambda_0 \cap \Lambda_1 = \phi$ and $\Lambda \equiv \Lambda_0 \cup \Lambda_1 \subset \mathcal{S}(n)$. In many applications, Σ is a function of an r-dimensional parameter λ, i.e., $\Sigma = \Sigma(\lambda)$, and the null and alternative hypotheses may be described in terms of λ. Typical examples are the problems of testing for serial correlation, heteroscedasticity, etc., which will be treated in detail in Chapter 6. Now, for a group leaving the problem invariant, choose $\mathcal{G} = R^k \times R_+$ which acts on y and (β, σ^2) as

$$\begin{aligned} g(y) &= cy + Xb \\ g(\beta, \sigma^2) &= (c\beta + b, c^2\sigma^2) \end{aligned} \tag{3.58}$$

for $g = (b, c) \in \mathcal{G}$. Further, choose a matrix $Z: n \times (n - k)$, such that

$$ZZ' = I - X(X'X)^{-1}X' \qquad \text{and} \qquad Z'Z = I_{n-k}, \tag{3.59}$$

and define

$$\begin{aligned} v &= (X'X)^{-1/2}X'y, \qquad w = Z'y \\ \eta &= (X'X)^{1/2}\beta \qquad \text{and} \qquad Q' = (X(X'X)^{-1/2}, Z). \end{aligned} \tag{3.60}$$

Then $Q \in \mathcal{O}(n)$. From (3.56) the density of $\tilde{y} = Qy = \begin{pmatrix} v \\ w \end{pmatrix}$ is given by

$$|Q\theta Q'|^{-1/2} q((\tilde{y} - \tilde{\eta})'[Q\theta Q']^{-1}(\tilde{y} - \tilde{\eta}))$$

with $\tilde{\eta} = \begin{pmatrix} \eta \\ 0 \end{pmatrix}$ and $\theta = \sigma^2\Sigma$. By Lemma 1.5 in Chapter 1, the marginal density of w is of the form

$$\bar{f}(w \mid \theta) = |Z'\theta Z|^{-1/2}\bar{q}(\psi(w:\theta)) \qquad \text{with} \qquad \psi(w:\theta) = w'[Z'\theta Z]^{-1}w, \tag{3.61}$$

where $\bar{q}$ depends on q and (n, k) but not on (β, θ). Note that in terms of (v, w), $w/\|w\|$ is clearly a maximal invariant, so that any invariant test is a measurable function of $w/\|w\|$. Hence to consider the nonnull robustness of an invariant test, it suffices to show that the nonnull distribution of $w/\|w\|$ is independent of $\bar{q}$ or q. Since in terms of w, $w/\|w\|$ is a maximal invariant under group R_+ acting on w by cw where $c \in R_+$, and since any maximal invariant is a bimeasurable bijection of the maximal invariant $\pi(w) = R_+ w$, we evaluate the probability ratio of $\pi(w) = R_+ w$ by applying Theorem 3.5 to the marginal density of w given in (3.61). First in Assumption 3.1, take $\mathcal{G}_1 = \bar{\mathcal{G}}_1 = R_+$ and $\mathcal{G}_2 = \{e_2\}$ so that R_+ is identified with $\mathcal{G}_1 \times \mathcal{G}_2$, where $(c, e_2)w \equiv cw$ for $(c, e_2) \in \mathcal{G}_1 \times \mathcal{G}_2$. Then the left-invariant measure $\nu_1(dc)$ on $\mathcal{G}_1 = \mathcal{G}_1 \times \mathcal{G}_2$ is also right-invariant, and so $\Delta_1(c) \equiv 1$. In addition, $\chi(c, e) = \chi_1(c)$, since $\chi_2(e) = 1$. Hence, Assumption 3.1 is satisfied with this choice of $\mathcal{G}_1$ and $\mathcal{G}_2$. Further, $\mathcal{G}_1$ acts transitively on the range of ψ in (3.61), because

$$\psi((c, e_2)w:\theta) = \psi(cw:\theta) = c^2\psi(w:\theta) = c^2\psi((e_1, e_2)w:\theta). \tag{3.62}$$

This verifies Assumption 3.2. Note here that $\chi_1(c) = c^{n-k}$, because $\chi_1(c)$ is the inverse of the Jacobian of $w \to cw$ (see Section 2.3 of Chapter 2). For $h(x_2, g_2, \theta : y_0)$ in Theorem 3.5, we choose $h(w, e_2, \theta : 1) = \psi(w:\theta)^{-1/2}$ from (3.51) and (3.62). Therefore it follows from Theorem 3.5 that for $\theta_i (i = 0, 1)$, the probability ratio of the maximal invariant $\pi = \pi(w)$ is given by $K(w \mid \theta_1)/K(w \mid \theta_0)$, where from (3.50) and (3.61)

$$K(w \mid \theta) = |Z'\theta Z|^{-1/2}\psi(w:\theta)^{-(n-k)/2}.$$

Consequently, we obtain the following theorem (Kariya and Sinha, 1985).

Theorem 3.6. *For given $\Sigma_i \in \Lambda_i (i = 0, 1)$, the probability ratio $R = dP^{\pi}_{\Sigma_1}/dP^{\pi}_{\Sigma_0}$ of the maximal invariant π, when evaluated at $\pi = \pi(w)$, is given by*

$$R = R(\pi(w)) = \left(\frac{|Z'\Sigma_1 Z|}{|Z'\Sigma_0 Z|}\right)^{-1/2}\left(\frac{w'[Z'\Sigma_1 Z]^{-1}w}{w'[Z'\Sigma_0 Z]^{-1}w}\right)^{-(n-k)/2} \tag{3.63}$$

Furthermore, the distribution $P^{\pi}_{\Sigma_0}$ is independent of $\bar{q}$ or q.

Proof. The first part has been proved above. To show the second part, note that $\pi = \pi(w)$ is a bimeasurable bijection of the maximal invariant $T = w/\|w\|$. Hence it suffices to show that the distribution of T under $\Sigma = \Sigma_0$ is independent of $\bar{q}$. But writing $T = \Sigma_0^{1/2}z/\|\Sigma_0^{1/2}z\|$ with $z = \Sigma_0^{-1/2}w$, T is a function of z, which is invariant under $z \to cz$ for $c \in R_+$, and the density of z is given by $\bar{f}(z \mid \sigma^2 I) = \sigma^{-(n-k)}\bar{q}(z'z/\sigma^2)$. Hence, by Corollary 3.2, the distribution of T is independent of $\bar{q}$. This completes the proof.

By this theorem, for testing $\Lambda_0 = \{\Sigma_0\}$ versus $\Lambda_1 = \{\Sigma_1\}$ in (3.57), the distribution of any test that is a function of $T = w/\|w\|$ is completely free from q and (β, σ^2) in (3.55)–(3.56), whether it may be the null distribution or nonnull distribution. This implies the nonnull robustness of an invariant test. In particular, since R in (3.63) is the density of $\pi = \pi(w)$, with respect to $P_{\Sigma_0}^{\pi}$, and since the null distribution $P_{\Sigma_0}^{\pi}$ of π is independent of q and (β, σ^2), by applying the Neyman–Pearson lemma to the density R, the test which rejects $\{\Sigma_0\}$ for small values of

$$S \equiv w'[Z'\Sigma_1 Z]^{-1}w/w'[Z'\Sigma_0 Z]^{-1}w \tag{3.64}$$

is UMPI (uniformly most powerful invariant) against $\{\Sigma_1\}$. The nonnull robustness of this UMPI test follows from the fact that S is a function of T or π, which in turn implies the optimality robustness of the test because the power function remains the same under any q and (β, σ^2).

So far our problem has been to test a simple null hypothesis, $\{\Sigma_0\}$, versus a simple alternative, $\{\Sigma_1\}$. However, in many applications often encountered, Σ_1 is of a certain structure, say $\Sigma_1 = \Sigma_1(\lambda)$, with unknown parameter λ though $\Sigma_0 = I$. Then S in (3.64) in general depends on λ; hence, it cannot be a test statistic. Some problems with such $\Sigma_1(\lambda)$ will be treated in Chapter 6. Here, as special cases, for the case of $X = 0$, we shall treat the problems of testing intraclass correlation, heteroscedasticity, and serial correlation.

Example 3.3. (Kariya and Eaton, 1977). Assume $X = 0$ in (3.55), so that from (3.59) $Z = I$, from (3.60) $w = y$, and the density of y is given by

$$f(y \mid \Omega) = \bar{f}(y \mid \Omega) = |\Omega|^{-1/2}q(y'\Omega^{-1}y). \tag{3.65}$$

Here we assume that Ω is of the form

$$\Omega = \sigma^2\Sigma(\rho) \equiv \sigma^2[\lambda_1(\rho)(I - M) + \lambda_2(\rho)M] \qquad (\sigma^2 > 0), \tag{3.66}$$

and consider the problem of testing $\rho = \rho_0$ versus $\rho \neq \rho_0$. In the above, $\lambda_i \equiv \lambda_i(\rho)$'s are functions of an unknown real parameter ρ defined on a set Υ, such that $\lambda_2(\rho) > \lambda_1(\rho) > 0$ for all $\rho \in \Upsilon$, except for the null point $\rho = \rho_0$, in which case $\lambda_1(\rho_0) = \lambda_2(\rho_0) = 1$, and M is an $n \times n$ idempotent matrix of rank m. Note that under the null hypothesis $H: \rho = \rho_0, \Omega = \sigma^2 I$ or $\Sigma(\rho_0) = I$. Now letting $\rho_1(\neq \rho_0)$, by the above argument, the test with critical region $S < k$ is UMPI for testing $\rho = \rho_0$ versus $\rho = \rho_1$ (or, equivalently, $\Sigma(\rho_0) = I$ versus $\Sigma(\rho_1) \equiv \Sigma_1$ $(\neq I)$) under the group $\mathscr{G} = R_+$ acting on y by cy, where $S = y'\Sigma_1^{-1}y/y'y$. But, from $\Sigma_1^{-1} = \lambda_1^{-1}(I - M) + \lambda_2^{-1}M$ with $\lambda_i = \lambda_i(\rho_1)$, we get

$$S = \lambda_1^{-1} + [\lambda_2^{-1} - \lambda_1^{-1}]S_0 \qquad \text{with} \qquad S_0 = y'My/y'y,$$

and $S < k$ is reduced to $S_0 > k'$, which is independent of the fixed ρ_1. Hence the test with critical region $S_0 > k'$ is UMPI for testing $\rho = \rho_0$ versus $\rho \neq \rho_0$ in (3.65) and (3.66), whatever q may be, and the null distribution of S_0 does not depend on q and σ^2. Further, because the nonnull distribution of S is independent of q and σ^2, the nonnull distribution of S_0 is independent of q and σ^2, though it depends on ρ through $\lambda_i = \lambda_i(\rho)$. Here we shall derive an explicit form of the density of the statistic

$$U \equiv (n - m)y'My/my'(I - M)y, \tag{3.67}$$

which is in one–one correspondence with S_0. Choose $Q \in \mathcal{O}(n)$ such that

$$Q'\Sigma^{-1}Q = \text{diag}\{\underbrace{\lambda_1^{-1}, \ldots, \lambda_1^{-1}}_{n-m}, \underbrace{\lambda_2^{-1}, \ldots, \lambda_2^{-1}}_{m}\},$$

and let $z = Qy$. Then by (3.65) the density of $z = (z_1, \ldots, z_n)'$ with $\sigma^2 = 1$ is given by

$$\lambda_1^{-(n-m)/2}\lambda_2^{-m/2}q\left(\sum_1^{n-m} \lambda_1^{-1}z_i^2 + \sum_{n-m+1}^{n} \lambda_2^{-1}z_i^2\right).$$

Hence, by Lemma 1.6 in Chapter 1,

$$\sum_{n-m+1}^{n} (\lambda_2^{-1/2}z_i)^2 \Bigg/ \left[\sum_1^{n-m} (\lambda_1^{-1/2}z_i)^2 + \sum_{n-m+1}^{n} (\lambda_2^{-1/2}z_i)^2\right] \sim Be\left(\frac{m}{2}, \frac{n-m}{2}\right),$$

which implies the distribution of

$$\tau U = \tau(n - m)y'My/my'(I - M)y = \tau(n - m) \sum_{n-m+1}^{n} z_i^2 \Bigg/ m \sum_1^{n-m} z_i^2$$

is $F(m, n - m)$, the F distribution with d.f. $(m, n - m)$ where $\tau = \lambda_1/\lambda_2 = \lambda_1(\rho)/\lambda_2(\rho)$. Therefore, the distribution of U is obtained simply by transforming τU into U. In addition, the UMPI critical region $S_0 > k'$ is equivalent to $U > k''$. Denoting this test by ϕ, its power function is given by

$$\pi(\phi, \rho) = P(\tau U > k''\tau) = \int_{\tau k''}^{\infty} F(u \mid m, n - m)\, du, \tag{3.68}$$

where $F(u \mid m, n-m)$ is the density of $F(m, n-m)$. The cut-off point k'' is obtained from (3.68) with $\tau = 1$, because $\lambda_1(\rho_0) = \lambda_2(\rho_0) = 1$.

A special case of $\Sigma(\rho)$ of the form (3.66) is the intraclass correlation structure $\Sigma = (\rho_{ij})$ where $\rho_{ii} = 1$ and $\rho_{ij} = \rho \geq 0$ for all $i \neq j$. In fact, it is expressed in the form of Σ in (3.66) with $\lambda_1 = (1-\rho)$, $\lambda_2 = (1 - \rho + n\rho)$ and $M = \underset{\sim}{1}(\underset{\sim}{1}'\underset{\sim}{1})^{-1}\underset{\sim}{1}'$, where $\underset{\sim}{1} = (1, \ldots, 1)' \in R^n$. Then it follows from the above result that the test ϕ with critical region $U > k$ is UMPI for testing $\rho = 0$ versus $\rho > 0$. Moreover, the null distribution of U is $F(1, n-1)$, and the power function of ϕ is evaluated as

$$(3.69) \qquad \pi(\phi, \rho) = \int_{\tau k}^{\infty} F(u \mid 1, n-1)\, du,$$

where $\tau = (1-\rho)/(1-\rho+n\rho)$. It is noted that this power function is common, whatever the density of y may be, so long as it is of the form (3.65).

Another example of a scale matrix Ω of the form (3.66) is the heteroscedastic matrix

$$(3.70) \qquad \Omega = \text{diag}\{\sigma_1^2, \ldots, \sigma_2^2, \sigma_2^2, \ldots, \sigma_2^2\} = \sigma_1^2[\lambda_1(I - M) + \lambda_2 M],$$

where $\lambda_1 = 1$, $\lambda_2 = 1 + \rho$ with $\rho = 1 - (\sigma_2^2/\sigma_1^2)$ and $M = \text{diag}\{0, \ldots, 0, \underbrace{1, \ldots, 1}_{m}\}$.

Then the problem of testing $\sigma_1^2 = \sigma_2^2$ versus $\sigma_1^2 > \sigma_2^2$ is equivalent to that of $\rho = 0$ versus $\rho > 0$ and $\lambda_2(\rho) > \lambda_1(\rho) > 0$ for $\rho > 0$. Hence from (3.67), the test ϕ based on $U = (n-m)y_2'y_2/my_1'y_1$ is UMPI under any density of the form (3.65) where $y' = (y_1', y_2')$ with $y_2 \colon m \times 1$, and the power function of this test is given by

$$(3.71) \qquad \pi(\phi, \rho) = \int_{\tau k}^{\infty} F(u \mid m, n-m)\, du,$$

where $\tau = 1/(1+\rho)$, whatever q may be.

This testing problem under a somewhat different distributional assumption about y is discussed in Chapter 8.

Example 3.4. Again assume $X = 0$ and that the density of y is given by (3.65). But here Ω is assumed to be of the form

$$(3.72) \qquad \Omega = \sigma^2 \Sigma(\rho) \equiv \sigma^2[\gamma_1(\rho)I + \gamma_2(\rho)A]^{-1},$$

where A is a known matrix and $\gamma_i \equiv \gamma_i(\rho)$'s are functions of a real parameter ρ defined on a set Υ such that Υ contains 0, $\gamma_1(0) = 1$ and $\gamma_2(0) = 0$, and $\gamma_i(\rho) > 0$ for all $\rho \neq 0$ in $\Upsilon (i = 1, 2)$. This specification of the scale matrix Ω often represents exact or approximate serial correlation models, possibly after subtraction of a location parameter, such as an AR(1) model (autoregressive model of order 1), a circularly serial correlation model, etc. (see Anderson, 1948, 1971; Durbin and Watson, 1950, 1951). Then by the argument following Theorem 3.2, the test which rejects for small values of $S = y'\Sigma(\rho_1)^{-1}y/y'y$ is UMPI for testing $\rho = 0$ versus $\rho = \rho_1$ ($\neq 0$) in Υ under any density of the form (3.65). But from (3.72)

$$S = \gamma_1(\rho_1) + \gamma_2(\rho_1)S_0 \qquad \text{with} \qquad S_0 = y'Ay/y'y,$$

and the UMPI critical region is equivalent to $S_0 < k$, which does not depend on the fixed ρ_1. Therefore the test ϕ with critical region $S_0 < k$ is UMPI for testing $\rho = 0$ versus $\rho > 0$ under any density in (3.65) with scale matrix (3.72). Furthermore, the nonnull distribution of S_0 depends only on ρ, and is independent of q and σ^2.

Application 3.9. *Testing Sphericity.* Let Z be an $n \times p$ matrix with density

$$f(Z \mid \alpha, \sigma^2\Sigma) = |\sigma^2\Sigma|^{-n/2} q(\operatorname{tr}(Z - \underset{\sim}{1}\alpha')'(Z - \underset{\sim}{1}\alpha')(\sigma^2\Sigma)^{-1}), \tag{3.73}$$

where $\underset{\sim}{1} = (1, \ldots, 1)' \in R^n$ and $\alpha \in R^p$, and consider testing $\Sigma = I$ versus $\Sigma \neq I$. Assuming $\alpha = 0$ without loss of generality, except for loss of one degree of freedom, this problem is left invariant under the group $\mathscr{G} = \mathscr{G}_1 \times \mathscr{G}_2 = R_+ \times \mathcal{O}(p)$ acting on the left of Z by $g(Z) = cZQ'$ for $g = (c, Q) \in \mathscr{G}$. An invariant measure on $\mathscr{G}$ is $\nu = \nu_1 \times \nu_2$, with $\nu_1(dc) = dc/c$ and ν_2 being the invariant probability measure on $\mathscr{G}_2$, and $\chi(c, Q) = \chi_1(c)\chi_2(Q)$ with $\chi_2(Q) \equiv 1$. Hence Assumption 3.1 is satisfied. Further putting $\psi(Z\colon \theta) = \operatorname{tr} Z'Z\theta^{-1}$ with $\theta = \sigma^2\Sigma$, the subgroup R_+ acts transitively on the range of ψ, the left-invariant measure ν_1 is also right-invariant so that $\Delta_1 \equiv 1$, and the multiplier $\chi_1(c)$ is the inverse of the Jacobian of $Z \to cZ$ so that $\chi_1(c) = c^{np}$. Thus Assumption 3.2 is also satisfied with $\bar{\mathscr{G}}_1 = \mathscr{G}_1$. Consequently, choosing $y_0 = 1$ in (3.51) or from $c^2\psi(ZQ'\colon \theta) = 1$, the K function corresponding to (3.50) is evaluated as

$$K(Z \mid \theta) = |\theta|^{-n/2} \int_{\mathcal{O}(p)} [\operatorname{tr} QSQ'\theta^{-1}]^{-np/2} \nu_2(dQ) \tag{3.74}$$

where $S = Z'Z$. Hence the following theorem follows from Theorem 3.5.

Theorem 3.7. (*Kariya and Sinha,* 1985). *For testing* $\mathrm{H}\colon \theta = \sigma^2 I$ *versus* $\mathrm{K}\colon \theta = \sigma^2\Sigma$, *the probability ratio* R *of the maximal invariant* π *in* (3.47), *evaluated at* $\pi = \pi(Z)$, *is given by*

$$R = R(\pi(Z)) = |\Sigma|^{-n/2} \int_{\mathcal{O}(p)} [1 + F]^{-np/2} \nu_2(dQ)$$

where $F = \operatorname{tr}(\Sigma_1^{-1} - I)QSQ'/\operatorname{tr} S$. *Furthermore, the null distribution of* π *is independent of* q *and* σ^2.

This problem will be further discussed in Section 6.2 of Chapter 6.

A different approach to the nonnull robustness of a statistic $t(X)$ when $X: n \times p \sim N(\mu, I_n \otimes \Sigma)$ is to modify Theorem 3.2 and Corollary 3.3. As in the case of the null robustness in Approach (1) of Section 3, let

$$(\mu, \Sigma) \in \Theta_1 \times \Lambda_1 \qquad \text{with} \quad 0 \in \Theta_1$$

be an alternative hypothesis where $\Theta_1 \subset R^{n \times p}$ and $\Lambda_1 \subset \mathscr{S}(p)$. Since Θ_1 and Λ_1 are arbitrary, this formulation includes the null case. Let $\mathscr{P}_E(\mu, I_n \otimes \Sigma)$ be the class of elliptical symmetric distributions with location parameter μ and scale matrix Σ. Recall that if a member in $\mathscr{P}_E(\mu, I \otimes \Sigma)$ has a density, it is of the form

$$f(X \mid \mu, \Sigma) = |\Sigma|^{-n/2} q(\operatorname{tr}(X - \mu)\Sigma^{-1}(X - \mu)').$$

Also recall that the nonnull robustness of a statistic $t(X)$ in $\mathscr{P}_E(\mu, I \otimes \Sigma)$, which is not necessarily a test statistic, means that for *each* $(\mu, \Sigma) \in \Theta_1 \times \Lambda_1$, $\mathscr{L}(t(X))$ under $\mathscr{L}(X) \in \mathscr{P}_E(\mu, I_n \otimes \Sigma)$ is the same as that under $\mathscr{L}(X) = N(\mu, I_n \otimes \Sigma)$.

Theorem 3.8. *The distribution of a statistic* $t(X)$ *is unique in* $\mathscr{P}_E(\mu, I_n \otimes \Sigma)$ *for each* $(\mu, \Sigma) \in \Theta_1 \times \Lambda_1$ *if the following conditions* (a) *and* (b) *hold*:

(a) $t(X - \mu) = t(X)$ *for all* $\mu \in \Theta_1$
(b) $t(\alpha X) = t(X)$ *for all* $\alpha > 0$.

Proof. Fix $(\mu, \Sigma) \in \Theta_1 \times \Lambda_1$. By (a), we can assume $\mu = 0$ without loss of generality, whereas from (b), $\mathscr{L}(t(X)) = \mathscr{L}(t(Y\Sigma^{1/2}) = \mathscr{L}(t(Y\Sigma^{1/2}/\|Y\|))$ where $Y = X\Sigma^{-1/2}$ and $\|Y\| = (\operatorname{tr} Y'Y)^{1/2}$. Since $\mathscr{L}(Y) \in \mathscr{P}_E(0, I_n \otimes I_p)$, $\mathscr{L}(Y/\|Y\|)$ is the unique uniform distribution over

$$\mathscr{U} = \{U: n \times p \mid \operatorname{tr} U'U = 1\}.$$

Hence $\mathscr{L}(t(X))$ remains the same in $\mathscr{P}_E(\mu, I_n \otimes \Sigma)$.

This result was obtained in Jensen and Good (1981).

We apply this result to some problems of testing on the structure of Σ, including Applications 3.1 and 3.2 given above. Using the same notation as in B of Section 3.2, let

$$X = (X_1', X_2', X_3')' \sim N(\mu, I \otimes \Sigma) \qquad \text{with} \qquad \mu = (\mu_1', \mu_2', \mu_3')'$$

be the MANOVA model, where $\mu_3 = 0$, and X_i: $n_i \times p$ $(i = 1, 2, 3)$, and consider an alternative hypothesis

$$(3.75) \qquad (\mu, \Sigma) \in \Theta_1 \times \Lambda_1 \qquad \text{with} \qquad \Theta_1 = R^{n_1 \times p} \times R^{n_2 \times p} \times \{0\},$$

where $\Lambda_1 \subset \mathscr{S}(p)$ is arbitrary. Then $(0, 0, 0) \in \Theta_0$ and invariant tests for the structure of Σ is based on $Z = X_3 \sim N(0, I_m \otimes \Sigma)$ where $m = n_3$ (see Section 3.2). Hence let $t(X)$ be a statistic based on Z only, which is not necessarily a test statistic. Then clearly (a) Theorem 3.8 is satisfied. Therefore, whenever $t(X)$ satisfies (b), the nonnull robustness of $t(X)$ in $\mathscr{P}_{\text{E}}(\mu, I \otimes \Sigma)$ for each $(\mu, \Sigma) \in \Theta_1 \otimes \Lambda_1$ holds. For example, in the problem of testing the sphericity H: $\Sigma = \sigma^2 I$ versus K: $\Sigma \neq \sigma^2 I$, treated in Application 3.9, $\Lambda_1 = \mathscr{S}(p) - \{\sigma^2 I\colon \sigma^2 > 0\}$. Hence letting $t(X) = Z/\|Z\|$, the nonnull robustness of $t(X)$ in $\mathscr{P}_{\text{E}}(\mu, I \otimes \Sigma)$ for each $(\mu, \Sigma) \in \Theta_1$ follows. This implies that any invariant test is nonnull robust since a maximal invariant is a function of $t(X)$.

Application 3.10. *Canonical Correlations and Test of Independence* (see Application 3.7). Let $S \equiv (S_{ij}) = Z'Z$, where $Z = (Z_1, Z_2) \sim N(0, I_m \otimes \Sigma)$ with Z_i: $m \times p_i$ and $S_{ij} = Z_i'Z_j (i, j = 1, 2)$. Let $t(X) = S_{11}^{-1} S_{12} S_{22}^{-1} S_{21}$ so that the canonical correlations are the squared latent roots of $t(X)$. Here, by choosing $\Lambda_1 = \mathscr{S}(p)$ in (3.75), we are considering the nonnull situation, including the null case in the problem of testing independence: $\Sigma_{12} = 0$ versus $\Sigma_{12} \neq 0$, where $\Sigma = (\Sigma_{ij})$ with Σ_{ij}: $p_i \times p_j$ $(i, j = 1, 2)$. Now obviously $t(\alpha X) = t(X)$ for all $\alpha > 0$, hence for each $(\mu, \Sigma) \in \Theta_1 \times \Lambda_1$ in (3.75), the nonnull robustness of $t(X)$ in $\mathscr{P}_{\text{E}}(\mu, I \otimes \Sigma)$ follows. In the null case where $\Sigma_{12} = 0$, by Corollary 3.3, the distribution of $t(X)$ is shown to be unique in $\mathscr{P}_{\text{E}} = \bigcup \{\mathscr{P}_{\text{E}}(\mu, I \otimes \Sigma) \,|\, (\mu, \Sigma) \in \Theta_0 \times \Lambda_0\}$ where $\Theta_0 = \Theta_1$, and $\Lambda_0 = \{\Sigma \in \mathscr{S}(p) \,|\, \Sigma_{12} = 0\}$. But in the nonnull case, where $\Sigma_{12} \neq 0$, the distribution of $t(X)$ certainly depends on Σ, and so the nonnull robustness holds for each fixed Σ. This implies the nonnull robustness of the canonical correlations and all the invariant tests for independence, because the set of the canonical correlations is a maximal invariant. This implies the optimality robustness of invariant tests, because the power function of each invariant test remains the same in $\mathscr{P}_{\text{E}}(\mu, I \otimes \Sigma)$. For example, the LBI test with critical region

$$T_4 \equiv \operatorname{tr} S_{11}^{-1} S_{12} S_{22}^{-1} S_{21} > c$$

under $\mathscr{L}(X) = N(\mu, I_n \otimes \Sigma)$, which is known as Pillai's test, remains LBI in $\mathscr{P}_{\text{E}}(\mu, I \otimes \Sigma)$. Similarly, the UMPI R^2-test in the case of $\min\{p_1, p_2\} = 1$, which is equivalent to Pillai's test, also remains UMPI in the above class.

When $p_1 = p_2 = 1$, the statistic $t(X)$ reduces to $r^2 = S_{12}^2/(S_{11}S_{22})$; the square of the sample correlation and naturally its distribution is the same for all $\mathscr{L}(X) \in \mathscr{P}_E(\mu, I \otimes \Sigma)$ for each $(\mu, \Sigma) \in \Theta_1 \times \mathscr{S}(p)$ with Θ_1 in (3.75). On the other hand, in Application 3.7, the null distribution of the sample correlation has been shown to remain the same in a broader class of distributions.

Application 3.11. *Tests for Outliers* (see D in Section 3.2). Let X be an $n \times p$ matrix with density

$$(3.76) \quad f(X \mid \alpha, \Delta, \Sigma) = |D_\Delta|^{-p/2}|\Sigma|^{-n/2}q(\operatorname{tr}\Sigma^{-1}(X - \underset{\sim}{1}\alpha')'D_\Delta^{-1}(X - \underset{\sim}{1}\alpha')),$$

where D_Δ is given in (3.22), and consider the problem of testing H: $\Delta = 0$ versus K: $\Delta > 0$. As in the previous application, assuming $\alpha = 0$ without essential loss of generality, this problem is left invariant under the group $\mathscr{G} = \mathscr{G}_1 \times \mathscr{G}_2 = \mathrm{GL}(p) \times \mathscr{P}$ acting on the left of X by $g(X) = \Gamma X C$ for $g = (C, \Gamma) \in \mathscr{G}$. Here $\mathscr{P}$ is the finite group of permutations of $\underset{\sim}{v}$. An invariant measure on $\mathscr{G}$ is $v = v_1 \times v_2$, with $v_1(dC) = dC/|C'C|^{p/2}$ and v_2 being the invariant probability measure on $\mathscr{P}$(assigning the mass $\dfrac{1}{n!}$ to each permutation $\underset{\sim}{v}$ in $\mathscr{P}$), and $\chi(C, \Gamma) = \chi_1(C) \cdot \chi_2(\Gamma)$ with $\chi_2(\Gamma) \equiv 1$. Hence Assumption 3.1 is satisfied. On the other hand, setting $\psi(X\colon \theta) = \operatorname{tr}\Sigma^{-1}X'D_\Delta^{-1}X$, it follows that $\psi(g(X)\colon \theta) = \operatorname{tr} C\Sigma^{-1}C'X'\Gamma'D_\Delta^{-1}\Gamma X$ satisfies (3.49) with $\bar{\mathscr{G}} = \mathscr{G}_1$, which acts transitively on the range space $\mathscr{Y}$ of ψ. Also, with the left-invariant measure $dC/|C'C|^{p/2}$ being also the right-invariant measure, $\Delta_1 \equiv 1$, and the multiplier $\chi_1(C)$ is the inverse of the Jacobian $X \to XC$, so that $\chi_1(C) = |C'C|^{n/2}$. Consequently, by choosing $y_0 = 1$ in (3.51), the K function corresponding to (3.50) is obtained as

$$(3.77) \quad \begin{aligned} K(X \mid \theta) &= |D_\Delta|^{-p/2}|\Sigma|^{-n/2}\int_{\mathscr{P}} |X'\Gamma'D_\Delta^{-1}\Gamma X|^{-n/2}|\Sigma|^{n/2}v_2(d\Gamma) \\ &= |D_\Delta|^{-p/2} \cdot \frac{1}{n!}\sum_\alpha |X'\Gamma_\alpha' D_\Delta^{-1}\Gamma_\alpha X|^{-n/2}, \end{aligned}$$

where $\sum_\alpha$ is summation over $n!$ permutations Γ_α of $\underset{\sim}{v}$, and the elements $a_{v_1}, \ldots, a_{v_n}$ appearing in D_Δ correspond to a fixed permutation, say $(1, \ldots, n)$.

The following result therefore follows from Theorem 3.5.

Theorem 3.9. *For testing* H: $\Delta = 0$ *versus* K: $\Delta > 0$ *in model* (3.76) *with* $\alpha = 0$, *the probability ratio R of the maximal invariant π in* (3.47), *evaluated at* $\pi = \pi(X)$, *is given by*

$$(3.78) \quad R = R(\pi(X)) = |D_\Delta|^{-p/2}\sum_\alpha \{|X'\Gamma_\alpha' D_\Delta^{-1}\Gamma_\alpha X|/|X'X|\}^{-n/2}.$$

Furthermore, an invariant null-robust test is also nonnull-robust, with distribution of the test statistic independent of q and Σ.

If $\alpha = 0$ is not assumed in (3.76), there is a third component $\mathcal{G}_3 = R^p$ in the product group $\mathcal{G}$ that now becomes $\mathcal{G} = \mathcal{G}_1 \times \mathcal{G}_2 \times \mathcal{G}_3$, with the group action on X as $X \to g(X) = \Gamma X C + \underset{\sim}{1}\eta'$ for $g = (C, \Gamma, \eta) \in \mathcal{G}$. An argument similar to the one used before to make Theorem 3.5 directly applicable leads to the following result. A direct proof is given in Section 7.3 of Chapter 7.

Theorem 3.10. (Das and Sinha, 1986). *For testing* H: $\Delta = 0$ *versus* K: $\Delta > 0$ *in the model* (3.76) *with* $\alpha \in R^p, \Sigma \in \mathcal{S}(p)$ *unknown, the probability ratio R of the maximal invariant* π *in* (3.47), *evaluated at* $\pi = \pi(X)$, *is given by*

$$(3.79) \qquad R = R(\pi(X)) = |D_\Delta|^{-p/2}\tau^{-p/2}(n^{p/2}/n!)\sum_{\alpha}\{|S_\alpha(\Delta)|/|S|\}^{-(n-1)/2},$$

where $\tau = \underset{\sim}{1}'D_\Delta^{-1}\underset{\sim}{1}$, $S = X'(I_n - \underset{\sim}{1}\underset{\sim}{1}'/n)X$, $\underset{\sim}{1} = (1, \ldots, 1)' \in R^n$, *and*

$$(3.80) \qquad S_\alpha(\Delta) = X'\Gamma_\alpha'(D_\Delta^{-1} - D_\Delta^{-1}\underset{\sim}{1}\underset{\sim}{1}'D_\Delta^{-1}/\tau)\Gamma_\alpha X, \qquad \Gamma_\alpha \in \mathcal{P}.$$

Moreover, an invariant null-robust test is also nonnull-robust.

This problem will be further discussed in Section 7.3 of Chapter 7.

3.5. General Approach to Optimality Robustness

Except in the situation described in Section 3.4, nonnull robustness seldom holds in general. This is true especially when a hypothesis on a location parameter is tested. In this section, we consider a general approach to the optimality robustness of UMPI (uniformly most powerful invariant) and LBI (locally best invariant) properties.

First, we consider the optimality robustness of the UMPI property. Adopting the framework and the notation used in Section 3.4, let $\mathcal{X}$ be a nonempty open subset of R^n, $\mathcal{F}(\Theta)$ the class of densities of the form

$$(3.81) \qquad f(x \mid \theta) = \beta(\theta)q(\psi(x:\theta)), \qquad \theta \in \Theta,$$

and $\mathcal{G}$ a locally compact and sigma-compact group acting on the left of $\mathcal{X}$, leaving invariant the problem

$$(3.82) \qquad \text{H}: \theta \in \Theta_0 \qquad \text{versus} \qquad \text{K}: \theta \in \Theta_1,$$

where $\psi(\cdot:\theta)$ is a known measurable function from $\mathcal{X}$ onto a subset $\mathcal{Y}$ of R^m and $\beta(\cdot)$ is a known function. Then as in (3.47), the probability ratio of the maximal invariant $\pi = \pi(x)$, the projection of $\mathscr{X}$ onto $\mathscr{X}/\mathscr{G}$, is given by

$$R = h(\pi(x)\,|\,\theta_1) = (dP^{\pi}_{\theta_1}/dP^{\pi}_{\theta_0})(\pi(x)) = H(x\,|\,\theta_1)/H(x\,|\,\theta_0) \tag{3.83}$$

with

$$H(x\,|\,\theta) = \int_{\mathscr{G}} f(gx\,|\,\theta)\chi(g)\nu(dg), \tag{3.84}$$

where χ is a left-multiplier function and ν is an invariant measure on $\mathscr{G}$ and $\theta_i \in \Theta_i$ $(i = 0, 1)$. We assume the null robustness of π or, equivalently, that $dP^{\pi}_{\theta_0}$ is independent of q in (3.81). Of course, if the nonnull robustness holds, as in the situation described in Section 3.4, the ratio in (3.83) does not depend on q, and thus the optimality robustness of an invariant test follows, whatever optimality may be concerned. In particular, the UMPI property of the R^2-test is optimality-robust (see Applications 3.7 and 3.10). Now, by the assumption of the null robustness of the maximal invariant π, the ratio $h(\pi(x)\,|\,\theta_1)$ is the density of π evaluated at $\pi = \pi(x)$ with respect to $dP^{\pi}_{\theta_0}$. Here, by abstracting the situations we will consider in the following chapters, we make the following assumption.

Assumption 3.3. (1) $H(x\,|\,\theta_0)$ is expressed as $k(x)K_0$, where K_0 is a constant depending on q but independent of $\theta_0 \in \Theta_0$ and $k(x) > 0$ (a.e.). (2) For each $\theta_1 \in \Theta_1$, $H(x\,|\,\theta_1)$ is expressed as a function of a real statistic $v = v(x)$ with $0 \leq v \leq 1$ in the form $H(x\,|\,\theta_1) = k(x)K(v(x)\,|\,\theta_1)$, such that $K(-v\,|\,\theta_1) = K(v\,|\,\theta_1)$ and $K(v\,|\,\theta_1)$ is convex in v, where $k(x)$ is the same as $k(x)$ in (1).

Theorem 3.11. *Under Assumption* 3.3, *the test with critical region* $v > c$ *is UMPI for testing the problem in* (3.82).

Proof. Since $h(\pi(x)\,|\,\theta_1)$ is the density of $\pi = \pi(x)$ with respect to $dP^{\pi}_{\theta_0}$, which is independent of q and $\theta_0 \in \Theta_0$, by applying the Neyman–Pearson lemma to the problem of testing θ_0 versus θ_1, a MPI test is given by the critical region

$$h(\pi(x)\,|\,\theta_1) > c\,h(\pi(x)\,|\,\theta_0) = c \tag{3.85}$$

where $h(\pi(x)\,|\,\theta_0) = 1$ is used. By (3.83) and Assumption 3.3, (3.85) is equivalent to

$$K(v\,|\,\theta_1) > c', \tag{3.86}$$

where $c' = cK_0$ is a constant that may depend on q. Here, by using (2) of Assumption 3.3, we obtain for $1 \geq \alpha \geq \frac{1}{2}$

$$K(v\,|\,\theta_1) = \alpha K(v\,|\,\theta_1) + (1-\alpha)K(-v\,|\,\theta_1) \geq K((2\alpha - 1)v\,|\,\theta_1).$$

This implies that $K(v\,|\,\theta_1)$ is a nondecreasing function of $v \in [0,1]$. Hence (3.86) yields the MPI critical region $v > c''$. Since it does not depend on θ_1, it is UMPI, completing the proof.

In applications, the convexity of $K(v\,|\,\theta_1)$ in v required in (2) of Assumption 3.3 is often guaranteed by the convexity of q, whereas the symmetry of $K(v\,|\,\theta_1)$ in v is proved by the invariance of the integral over $\mathscr{G}$ in $H(x\,|\,\theta_1)$. In Chapter 5, the UMPI properties of the F-test in the ANOVA, Hotelling's T^2-test and the LRT in a special case of the MANOVA are shown to be optimality-robust by the approach in Theorem 3.11.

Example 3.5. (see Example 3.1). Let x have a density $f(x\,|\,\mu,\sigma^2)$ in $\mathscr{F}_{\mathrm{E}}(\mu \underset{\sim}{1}, \sigma^2 I_n)$ and consider the Student's t-problem H: $\mu = 0$ versus K: $\mu \neq 0$, where $\mathscr{F}_{\mathrm{E}}(\mu \underset{\sim}{1}, \sigma^2 I_n)$ is the class of densities of the form

$$(3.87) \qquad f(x\,|\,\mu,\sigma^2) = \sigma^{-n} q(\|x - \mu \underset{\sim}{1}\|^2/\sigma^2) \qquad \text{with } q \text{ convex on } [0,\infty),$$

and $\underset{\sim}{1} = (1,\ldots,1)' \in R^n$. Here, in comparison with (3.81), $\theta = (\mu,\sigma^2)$, $\beta(\theta) = \sigma^{-n}$ and $\psi(x\,|\,\theta) = \|x - \mu \underset{\sim}{1}\|^2/\sigma^2$. Then the problem remains invariant under the group $\mathscr{G} = R_* = \{a \in R\,|\,a \neq 0\}$ acting on x by $x \to ax$ where $a \in \mathscr{G}$. Of course, $\mathscr{G}$ is locally compact and sigma-compact, and its action on x is proper. Further $\chi(g) = |a|^n$ and $v(dg) = da/|a|$ (see Chapter 2). Hence from (3.83), the probability ratio of the maximal invariant π is given by

$$h(\pi(x)\,|\,\mu_1,\sigma_1^2) = H(x\,|\,\mu_1,\sigma_1^2)/H(x\,|\,0,\sigma_0^2)$$

where

$$(3.88) \qquad H(x\,|\,\mu,\sigma^2) = \int_{-\infty}^{\infty} \sigma^{-n} q(\|ax - \mu \underset{\sim}{1}\|^2/\sigma^2)|a|^{n-1}\,da,$$

and $\mu_1 \neq 0$, σ_0^2 and σ_1^2 are arbitrarily fixed. Here, because $\pi = \pi(x)$ is in one–one correspondence with another maximal invariant $u(x) = x/\|x\|$ and because the null distribution of $u(x)$ is the uniform distribution on $\mathscr{U} = \{u \in R^n\,|\,\|u\| = 1\}$ independently of q (see Corollary 3.3 or Section 1.3 of Chapter 1), the null distribution of π is independent of q. To verify Assumption 3.3, first consider $H(x\,|\,0,\sigma_0^2)$. From (3.88), it is easily shown to be $\|x\|^{-n} K_0$ with $K_0 = \int q(a^2)|a|^{n-1}\,da$. Hence, (1) in Assumption 3.3 is satisfied. Next we evaluate $H(x\,|\,\mu_1,\sigma_1^2)$. From (3.88), it is expressed as

$$(3.89) \qquad H(x\,|\,\mu_1,\sigma_1^2) = \int_{-\infty}^{\infty} \sigma_1^{-n} q[(a^2\|x\|^2 - 2a\mu_1 x' \underset{\sim}{1} + n\mu_1^2)/\sigma_1^2]|a|^{n-1}\,da.$$

Here, transforming a into $b = a\|x\|/\sigma_1$ and b into $-b$ yields

(3.90)
$$H(x \mid \mu_1, \sigma_1^2) = \|x\|^{-n} \int_{-\infty}^{\infty} q\left[b^2 - 2b\sqrt{n}\frac{\mu_1|v|}{\sigma_1} + \frac{n\mu_1^2}{\sigma_1^2}\right]|b|^{n-1}\,db$$
$$\equiv \|x\|^{-n} K(|v| \mid \mu_1, \sigma_1^2),$$

with

(3.91)
$$v = x'\underset{\sim}{1}/\|x\|\,\|\underset{\sim}{1}\|.$$

Here clearly $0 \leq v \leq 1$, and $K(-|v| \mid \mu_1, \sigma_1^2) = K(|v| \mid \mu_1, \sigma_1^2)$, because in the integral of (3.90), changing b into $-b$ leaves the integral the same. Further the convexity of $K(|v| \mid \mu_1, \sigma_1^2)$ follows from the convexity of q (Exercise 3.7). This implies that (2) in Assumption 3.3 is verified. Therefore, noting that v in (3.91) and $t = \sqrt{n}\bar{x}/s$ (see Example 3.1) are related as

(3.92)
$$t = \sqrt{n-1}\,v/\sqrt{1-v^2},$$

the present example is closed with

Theorem 3.12. *The t-test with critical region $|t| > c$ is UMPI in the class of densities of the form* (3.87), *whatever q may be, so long as it is convex.*

Next, we consider the robustness of an LBI property. In the same framework as in the case of the robustness of the UMPI property, assume $\Theta_0 = \{\theta_0\}$, a simple null hypothesis in (3.82) and that $\psi(x\colon \theta)$ in (3.81) is a function from $\mathscr{X}$ into $\mathscr{Y} = [0, \infty)$. Further it is assumed that $\{\theta_0\} \cup \Theta_1$ is open and $\theta_0 \in \bar{\Theta}_1$ where $\bar{\Theta}_1$ is the closure of Θ_1, and that q and β are continuously twice differentiable. With these assumptions, we expand the integrand of the numerator $H(x \mid \theta_1)$ in the probability ratio of the maximal invariant $\pi = \pi(x)$ of (3.83) as

(3.93)
$$f(x \mid \theta_1) = \beta(\theta_1)\{q(\psi(x\colon \theta_0)) + q^{(1)}(\psi(x\colon \theta_0))[\psi(x\colon \theta_1) - \psi(x\colon \theta_0)]$$
$$+ \frac{1}{2} q^{(2)}(\psi^*(x\colon \theta_0, \theta_1))[\psi(x\colon \theta_1) - \psi(x\colon \theta_0)]^2\}$$

where $\beta(\theta_1) = \beta(\theta_0) + \kappa + o_1$ with $\kappa = O(\|\theta_1 - \theta_0\|)$ and $o_1 = o(\|\theta_1 - \theta_0\|)$, $\psi^*(x\colon \theta_0, \theta_1) = c\psi(x\colon \theta_0) + (1-c)\psi(x\colon \theta_1)$ for some $0 \leq c \leq 1$ and $q^{(i)} = \partial^i q/\partial z^i$. Then, with $D = H(x \mid \theta_0)$, the ratio R in (3.83) is expressed as

(3.94)
$$R = 1 + J(x\colon \theta_0, \theta_1) + M(x\colon \theta_0, \theta_1),$$

where

$$J(x: \theta_0, \theta_1)$$
$$= (\beta(\theta_0) + \kappa) \int_{\mathscr{G}} q^{(1)}(\psi(gx: \theta_0))[\psi(gx: \theta_1) - \psi(gx: \theta_0)]\chi(g)v(dg)/D,$$

and $M(x: \theta_0, \theta_1)$ is a remainder term. Here if we can show that $J(x: \theta_0, \theta_1)$ is expressed as $\gamma(\theta_0, \theta_1)s(x) + \lambda(\theta_0, \theta_1)$ with $\gamma(\theta_0, \theta_1) = O(\|\theta_1 - \theta_0\|)$ and that for any invariant test $\phi(x)$ of size α,

$$\int \phi(x) M(x: \theta_0, \theta_1)\, dP^{\pi}_{\theta_0} = o(\|\theta_1 - \theta_0\|) \qquad \text{uniformly in } \phi,$$

then from (3.94) the power function is given by

$$\pi(\phi, \theta_1) = \alpha + \gamma(\theta_0, \theta_1) E_{\theta_0}[\phi(x)s(x)] + \alpha\lambda(\theta_0, \theta_1) + o(\|\theta_1 - \theta_0\|) \tag{3.95}$$

for $\theta_1 \in \Theta_1$ in a neighborhood of θ_0. Hence maximizing $E_{\theta_0}[\phi(x)s(x)]$ by using the generalized Neyman–Pearson lemma, the test based on $s(x)$ is LBI. Sometimes the second term inside of { } in (3.93) vanishes, in which case higher order derivatives of q need to be considered. In this manner, most LBI tests can be derived. A typical example is the LBI test in the GMANONA problem treated in Chapter 5. Here, as an example, we shall derive an LBI test for serial correlation in regression.

Example 3.6. Serial Correlation in Regression (see C in Section 3.2 and Application 3.8). Let $y = X\beta + u$ be a regression model with error term u, having the density

$$f(u \mid \sigma^2\Sigma) = |\sigma^2\Sigma|^{-1/2} q(u'\Sigma^{-1}u/\sigma^2), \tag{3.96}$$

where $X: n \times k$ is a fixed matrix of rank k,

$$\Sigma = \Sigma(\rho) = (\gamma_{ij}) = (I - \rho A + \rho^2 B)^{-1} \tag{3.97}$$

with $\gamma_{ij} = \rho^{|i-j|}/(1 - \rho^2)$, and A and B are defined in (3.20). When f in (3.96) is a normal density, the scale matrix $\Sigma(\rho)$ in (3.97) corresponds to the AR(1) process of errors: $u_i = \rho u_{i-1} + \varepsilon_i$. The problem we consider is to test H: $\rho = 0$ versus K: $\rho > 0$ or, equivalently,

$$\text{H: } \Sigma(0) = I \qquad \text{versus} \qquad \text{K: } \Sigma = \Sigma(\rho), \qquad \rho > 0.$$

As discussed in Application 3.8, the problem remains invariant under the group $\mathscr{G} = R^k \times R_+$ acting on y by (3.58). With $Z: n \times (n - k)$ in (3.59) and $w = Z'y$ in (3.60), respectively, it follows from Theorem 3.6 that the density of the maximal invariant $\pi = \pi(w)$ with respect to dP_0^π is given by

$$h(\pi(w)\,|\,\rho) = |Z'\Sigma(\rho)Z|^{-1/2}\{w'[Z'\Sigma(\rho)Z]^{-1}w/w'w\}^{-(n-k)/2} \tag{3.98}$$

where P_0^π is the distribution of π under $\rho = 0$. This is an expression of (3.83). To expand h in (3.98) in a neighborhood of $\rho = 0$, note that from $Z'Z = I_{n-k}$

$$Z'\Sigma(\rho)Z = I_{n-k} + \rho Z'AZ + O(\rho^2).$$

By using this, the density is expanded as

$$\begin{aligned} h(\pi(w)\,|\,\rho) &= [1 - \tfrac{1}{2}\rho \operatorname{tr} Z'AZ + O(\rho^2)]\left[1 + \frac{n-k}{2}\rho d + O\left(\rho^2, \frac{w}{\|w\|}\right)\right] \\ &= 1 + \frac{n-k}{2}\rho d - \tfrac{1}{2}\rho \operatorname{tr} Z'AZ + O\left(\rho^2, \frac{w}{\|w\|}\right) \end{aligned} \tag{3.99}$$

with

$$d = w'Z'AZw/w'w = e'Ae/e'e, \tag{3.100}$$

where $e = [I - X(X'X)^{-1}X']y \equiv Ny$. In a neighborhood of $\rho = 0$, $O(\rho^2, w/\|w\|)$ is shown to be $O(\rho^2)$ uniformly in w. Thus the power function of an invariant test of size α is locally expressed as

$$\pi(\phi, \rho) = \alpha + \frac{n-k}{2}\rho E_0[\phi d] - \frac{1}{2}\rho\alpha \operatorname{tr} NA + O(\rho^2). \tag{3.101}$$

Consequently, we obtain

Theorem 3.13. *For testing* H: $\rho = 0$ *versus* K: $\rho > 0$ *in the density* (3.96) *with* Σ *in* (3.97), *the test rejecting* H *for large values of* d *in* (3.100) *is LBI.*

Following the developments in this section, the robust LBI test for the GMANOVA hypothesis is derived in Chapter 5 and the robust LBI property of the $b_{2,p}$-test for detection of outliers under the model (3.21) or (3.22) is established in Chapter 7.

Exercises

3.1. Show that for testing H: $\mu_1 = 0$ versus K: $\mu_1 \neq 0$ in the model (3.10), a maximal invariant under the group $\mathscr{G}$ in (3.11) is the set of the ordered nonzero latent roots of $X_1'X_1(X_3'X_3)^{-1}$.

3.2. Show that for testing H: $\Sigma_{12} = 0$ versus K: $\Sigma_{12} \neq 0$ in the model (3.14), a maximal invariant under the group $\mathcal{G} = \mathcal{O}(m) \times \mathrm{GL}(p_1) \times \mathrm{GL}(p_2)$ is the set of the ordered nonzero latent roots of W.

3.3. Show that for testing H: $\Sigma = \sigma^2 I$ versus K: $\Sigma \neq \sigma^2 I$ in the model (3.14), a maximal invariant under the group $\mathcal{G} = \mathcal{O}(m) \times \mathcal{O}(p) \times R_+$ is $\{r_j/\Sigma_1^p r_j,\ j = 1,\ldots,p\}$ where $r_1 \geqq \cdots \geqq r_p$ are the ordered latent roots of S.

3.4. Show that for testing H: $\Sigma = I$ versus K: $\Sigma \neq I$ in the model (3.14), a maximal invariant is the set of ordered latent roots of S.

3.5. Show that the actions in (B1), (B2), (B3), (C) and (D) satisfy Cartan $\mathcal{G}$-space condition. Hint: see Example 2.2.

3.6. Show that the actions in (B1), (B2), (B3), (C) and (D) are proper. Hint: see Wijsman (1985).

3.7. Show that the function $K(|v| \,|\, \mu_1, \sigma_1^2)$ in (3.90) is convex when q is convex.

Chapter 4 Robustness of t-Test and Tests for Serial Correlation

4.1. Formulation of the Problem

In this chapter we consider the optimality robustness of the Student's t-test and tests for serial correlation, mainly without invariance. The null robustness and the robustness of the UMPI property have been shown in Chapter 3. First let us review some results on the optimalities of the t-test under normality. As in Example 3.5 of Chapter 3, for $z = (z_1, \ldots, z_n)'$, the t-test statistic

$$(4.1) \qquad t = \sqrt{n}\,\bar{z}/s \qquad \text{with} \quad \bar{z} = \Sigma z_i/n \quad \text{and} \quad s^2 = \Sigma(z_i - \bar{z})^2/(n-1)$$

is in one–one correspondence with

$$(4.2) \qquad v = z'\underset{\sim}{1}/\|z\|\,\|\underset{\sim}{1}\|.$$

Define the one-sided t-test and the two-sided t-test, respectively, by

$$(4.3) \qquad \phi_1(z) = \begin{cases} 1 & \text{if} \quad v > c \quad (\text{or} \quad t > c') \\ 0 & \text{otherwise} \end{cases}$$

and

$$(4.4) \qquad \phi_2(z) = \begin{cases} 1 & \text{if} \quad |v| > c \quad (\text{or} \quad |t| > c') \\ 0 & \text{otherwise.} \end{cases}$$

The following well-known results under normality, which we shall refer to frequently, are given without proofs.

Theorem 4.1. *Let* $z \sim N(\mu \underset{\sim}{1}, \sigma^2 I_n)$ *and let* $0 < \alpha < 1$ *be a significance level.*

(1) (*Lehmann and Stein,* 1948). *For* $1 > \alpha \geq 1/2$, *the one-sided t-test* ϕ_1 *is UMP for testing* $\mu = 0$ *versus* $\mu > 0$.

(2) (*Lehmann and Stein,* 1948). *For* $1/2 > \alpha > 0$, *no UMP test exists for testing* $\mu = 0$ *versus* $\mu > 0$.

(3) ϕ_1 *is UMPI for testing* $\mu = 0$ *versus* $\mu > 0$.

(4) ϕ_2 *is UMP similar and UMPI for testing* $\mu = 0$ *versus* $\mu \neq 0$.

Our goal is to study the robustness of these optimality properties of ϕ_1 and ϕ_2 in the class of spherically symmetric distributions whose densities are of the form

$$f(z \mid \mu, \sigma^2) = \sigma^{-n} q(\|z - \mu \underset{\sim}{1}\|^2/\sigma^2).$$

Here, since q and σ^2 are unknown, σ^2 can be absorbed into q without loss of generality when q is treated like a parameter. Understanding this point, let

$$\mathscr{Q} = \left\{ q \mid q\colon [0, \infty) \to [0, \infty), \int_{R^n} q(\|x\|^2)\, dx = 1 \right\} \tag{4.5}$$

and for a fixed vector $a \in R^n$ and for $\mu \in R$, define the three classes of densities on R^n:

$$\mathscr{F}_0(\mu) = \{f \mid f(x) = q(\|x - \mu \underset{\sim}{1}\|^2), \quad q \in \mathscr{Q}\}, \tag{4.6}$$

$$\mathscr{F}_1(\mu) = \{f \mid f(x) = q(\|x - \mu \underset{\sim}{1}\|^2), \quad q \in \mathscr{Q}, \quad q \text{ is nonincreasing}\}, \tag{4.7}$$

and

$$\mathscr{F}_2(\mu) = \{f \mid f(x) = q(\|x - \mu \underset{\sim}{1}\|^2), \quad q \in \mathscr{Q}, \quad q \text{ is nonincreasing and convex}\}. \tag{4.8}$$

Clearly $\mathscr{F}_2(\mu) \subset \mathscr{F}_1(\mu) \subset \mathscr{F}_0(\mu)$ for all $\mu \in R$. Let z be an $n \times 1$ random vector with density h and consider the testing problems

$$\mathrm{H}_0\colon h \in \mathscr{F}_0(0) \quad \text{versus} \quad \mathrm{K}_0\colon h \in \mathscr{F}_0(\mu), \quad \mu > 0. \tag{4.9}$$

$$\mathrm{H}_1\colon h \in \mathscr{F}_1(0) \quad \text{versus} \quad \mathrm{K}_1\colon h \in \mathscr{F}_1(\mu), \quad \mu > 0, \tag{4.10}$$

$$\mathrm{H}_2\colon h \in \mathscr{F}_2(0) \quad \text{versus} \quad \mathrm{K}_2\colon h \in \mathscr{F}_2(\mu), \quad \mu \neq 0. \tag{4.11}$$

Further, let

$$\mathscr{C}_i = \{\phi \in \mathscr{C} \mid E_f[\phi] \leq \alpha \quad \text{for all} \quad f \in \mathscr{F}_i(0)\} \tag{4.12}$$

be the class of level α tests under $H_i (i = 0, 1, 2)$ where $\mathscr{C}$ is the set of all tests on R^n, $0 < \alpha < 1$, and $E_f[\phi] \equiv E[\phi \mid f]$ denotes the expectation of ϕ under f. It is noted that $\mathscr{F}_2(0) \subset \mathscr{F}_1(0) \subset \mathscr{F}_0(0)$ implies $\mathscr{C}_0 \subset \mathscr{C}_1 \subset \mathscr{C}_2$. This fact is important for understanding the following results on optimality robustness.

Theorem 4.2. (1) (Lehmann and Stein, 1949). *For testing* H_0 *versus* h *of* $N(\mu \underset{\sim}{1}, \sigma^2 I_n)$ *with* $\mu > 0$, ϕ_1 *is UMP.*
(2) (Kariya and Eaton, 1977). *For testing* H_0 *versus* K_1, ϕ_1 *is UMP.*
(3) (Kariya and Eaton, 1977). *For testing* H_0 *versus* K_2, ϕ_2 *is UMP similar.*

Clearly (1) is a special case of (2). The proofs of (2) and (3) will be given for stronger results in Sections 4.2 and 4.3.

The result (1) above might apparently seem to contradict the nonexistence of a UMP test in (2) of Theorem 4.1 when $0 < \alpha < 1/2$. However, in (1) of Theorem 4.2 the class of level α tests for which ϕ_1 claims the UMP property is $\mathscr{C}_0$, but not $\mathscr{C}_N = \bigcap_{\sigma^2 > 0} \mathscr{C}_N(\sigma^2)$ with

$$\mathscr{C}_N(\sigma^2) = \{\phi \in \mathscr{C} \mid E[\phi \mid N(0, \sigma^2 I)] \leq \alpha\}, \tag{4.13}$$

and $\mathscr{C}_N$ is much bigger than $\mathscr{C}_0$. Also in the results of (2) and (3) in Theorem 4.2, the class for which the optimality properties hold is only $\mathscr{C}_0$. In fact, the UMP property of ϕ_1 in (2) simply means that for any $f(\cdot \mid \mu) \in \mathscr{F}_1(\mu)$ and $\mu > 0$,

$$E[\phi_1 \mid f(\cdot \mid \mu)] \geq E[\phi \mid f(\cdot \mid \mu)] \qquad \text{for all} \quad \phi \in \mathscr{C}_0, \tag{4.14}$$

but it does not mean that for a fixed $h(\cdot \mid \mu) \in \mathscr{F}_1(\mu)$, ϕ_1 is UMP for testing

$$H_1(h)\colon \mu = 0 \qquad \text{versus} \qquad K_1(h)\colon \mu > 0 \tag{4.15}$$

in the class of level α tests under the fixed h defined as

$$\mathscr{C}(h) = \{\phi \in \mathscr{C} \mid E[\phi \mid h(\cdot \mid 0)] \leq \alpha\}. \tag{4.16}$$

Of course, $\mathscr{C}(h)$ contains $\mathscr{C}_0$ for any fixed h. For example, consider the problem of testing $\mu = 0$ versus $\mu > 0$ in the normal model $N(\mu \underset{\sim}{1}, I)$. Then by the Neyman–Pearson lemma, the test ϕ^* with critical region $z' \underset{\sim}{1} > c$ is UMP in the class $\mathscr{C}_N(1)$. Clearly this UMP test ϕ^* is different from the UMP test ϕ_1 for H_0 versus K_1, although the density of $N(\mu \underset{\sim}{1}, I)$ belongs to $\mathscr{F}_1(\mu)$ in K_1. This implies that the power of ϕ^* is uniformly bigger than the power of ϕ_1 when the normal model $N(\mu \underset{\sim}{1}, I)$ is concerned. But ϕ^* does not belong to $\mathscr{C}_0$. In fact, by taking $N(0, \sigma^2 I)$ as a member belonging to $\mathscr{F}_0(0)$, it is easily shown that $E[\phi^* \mid N(0, \sigma^2 I)] > \alpha$ for $\sigma^2 < 1$. Hence $\mathscr{C}_0$ is a proper subset of $\mathscr{C}_N(1)$, and the fact that ϕ_1 is UMP in $\mathscr{C}_0$ does not contradict the fact that ϕ^* is UMP in $\mathscr{C}_N(1)$. Here it is remarked that the UMP-ness of ϕ^* follows from the information that the variance is known and hence the test ϕ^* need not have the structure of scale invariance like ϕ_1.

The above argument reveals that in the consideration of optimality robustness, it is extremely important to pay attention to the class in which the optimality holds. In Section 4.2, for each fixed $h \in \mathscr{F}_1(\mu)$, the one-sided test ϕ_1 is shown to be UMP for testing $\mu = 0$ versus $\mu > 0$, (i.e., for the problem 4.15) in the class of conditional level α tests given $W = \|z\|^2$, which is a subclass of $\mathscr{C}(h)$ in (4.16). Also the class $\mathscr{C}_0$ is identified with the intersection of the classes of conditional level α tests over $h \in \mathscr{F}_1(0)$. Hence these results imply (2) in Theorem 4.2. Further when $1/2 < \alpha < 1$, ϕ_1 is shown to be UMP in $\mathscr{C} = \bigcap_{h \in \mathscr{F}_1(0)} \mathscr{C}(h)$. In Section 4.3, for each fixed $h \in \mathscr{F}_2(\mu)$, the two-sided test ϕ_2 is shown to be UMP for testing $\mu = 0$ versus $\mu \neq 0$ in the class of conditional similar size α tests satisfying a certain condition. This result is also stronger than the result (3) in Theorem 4.2. In Section 4.4 the problem is treated with invariance. Finally in Section 4.5, the optimality robustness of the tests for serial correlation is considered without invariance.

This section ends with the following lemma, which will be used in Sections 4.2 and 4.3.

Lemma 4.1. *Let $v = z'\underset{\sim}{1}/\|z\|\,\|\underset{\sim}{1}\|$ and $w = \|z\|^2$, and let $h \in \mathscr{F}_0(\mu)$ be the density of z, where $h(z) = q(\|z - \mu\underset{\sim}{1}\|^2)$. Then*

(1) *(v, w) is a sufficient statistic for h.*
(2) *The joint density of (v, w) is given by*

$$g(v, w\colon \mu) = c_0 q(w - 2\sqrt{w}\, v \sqrt{n}\, \mu + n\mu^2) r_0(v) w^{n/2-1} \quad \text{with} \tag{4.17}$$

$$r_0(v) = 2\left[B\left(\frac{1}{2}, \frac{n-3}{2}\right)\right]^{-1} (1 - v^2)^{(n-3)/2} \tag{4.18}$$

where $c_0 = \Gamma(\frac{1}{2})^n / \Gamma(\frac{n}{2})$ and $B(a, b)$ denotes the beta function.

(3) *When $\mu = 0$, v and w are independent with density $r_0(v) r_1(w)$ where*

$$r_1(w) = c_0 w^{n/2-1} q(w). \tag{4.19}$$

Proof. (1) and (3) follow from (2) and the proof of (2). To show (2), transforming z_i's in $h(z) = q(\|z - \mu\underset{\sim}{1}\|^2)$ into the polar coordinates $z_i =$

$\sqrt{w}\,s_i(\theta)$ $(i = 1, \ldots, n)$, where

$$s_i(\theta) = \cos\theta_1 \cos\theta_2 \cdots \cos\theta_{i-1} \sin\theta_i \qquad (i = 1, \ldots, n-1)$$

$$s_n(\theta) = \cos\theta_1 \cos\theta_2 \cdots \cos\theta_{n-1},$$

$$\left(-\frac{\pi}{2} < \theta_i < \frac{\pi}{2},\ i = 1, \ldots, n-2,\ -\pi < \theta_{n-1} < \pi\right)$$

the joint density of $(w, \theta_1, \ldots, \theta_{n-1})$ is given by

$$c_0 q(w - 2\sqrt{w}\,v(\theta)\sqrt{n}\,\mu + n\mu^2) w^{n/2-1} J(\theta)$$

where $v(\theta) = \Sigma_{i=1}^{n} s_i(\theta)/\sqrt{n}$ and $J(\theta) = \cos^{n-2}\theta_1 \cos^{n-3}\theta_2 \cdots \cos\theta_{n-1}$. Here noting that $v = v(\theta)$, further transform $(w, \theta_1, \ldots, \theta_{n-1})$ into $(w, \theta_1, \ldots, \theta_{n-2}, v)$ for each region of $\theta_{n-1} \in (-\pi, -\frac{\pi}{2})$, $\theta_{n-1} \in [-\frac{\pi}{2}, 0)$, $\theta_{n-1} \in [0, \frac{\pi}{2})$, and $\theta_{n-1} \in [\frac{\pi}{2}, \pi)$. Then the joint density of $(w, \theta_1, \ldots, \theta_{n-2}, v)$ becomes

$$c_0 q(w - 2\sqrt{w}\,v\sqrt{n}\,\mu + n\mu^2) w^{n/2-1} R(v, \theta_1, \ldots, \theta_{n-2}),$$

where $R(v, \theta_1, \ldots, \theta_{n-2})$ does not depend on w. Hence, by replacing $R(v, \theta_1, \ldots, \theta_{n-2})$ by $r(v) = \int \cdots \int R(v, \theta_1, \ldots, \theta_{n-2})\, d\theta_1 \cdots d\theta_{n-2}$ yields the joint density of (v, w). To identify $r(v)$ with $r_0(v)$ in (4.18), set $\mu = 0$. Then $r(v)$ is the density of v, which is uniquely determined from the uniform distribution of $z/\|z\|$ on the sphere of R^n. Equivalently, it is determined from the t-distribution with degrees of freedom $n - 1$ through the one–one correspondence between t in (4.1) and v, and then it is identified with $r_0(v)$ in (4.18). This completes the proof.

By (1) of this lemma, without loss of generality we can restrict the class of tests to the one based on the sufficient statistic (v, w). We shall denote this class by $\mathscr{D}$.

4.2. One-Sided Testing Problems without Invariance

Let $z = (z_1, \ldots, z_n)'$ be a random vector with density h. In this section, we consider the following one-sided testing problems:

[i] For a fixed $h \in \mathscr{F}_1(\mu)$, $\mu = 0$ versus $\mu > 0$.
[ii] H_0: $h \in \mathscr{F}_0(0)$ versus K_1: $h \in \mathscr{F}_1(\mu)$, $\mu > 0$.
[iii] H_1: $h \in \mathscr{F}_1(0)$ versus K_1: $h \in \mathscr{F}_1(\mu)$, $\mu > 0$.

The $\mathscr{F}_i(\mu)$'s are defined in Section 4.1, and as has been discussed, the differences in the null hypotheses of these problems result in the differences of the classes of level α tests under the null hypotheses. For Problem [i], the class of level α tests is given by $\mathscr{C}(h)$ in (4.16), whereas for Problems [ii] and [iii] the classes are given by $\mathscr{C}_0$ and $\mathscr{C}_1$, respectively, given in (4.12). By Lemma 4.1, we can restrict each of the classes $\mathscr{C}_i$ and $\mathscr{C}(h)$ to the one based on the sufficient statistic $(v, w) = (v(z), w(z)) = (z'\underset{\sim}{1}/\|z\|\,\|\underset{\sim}{1}\|, \|z\|^2)$:

$$(4.20)\qquad \begin{cases} \mathscr{D}(h) = \{\psi \in \mathscr{D} \mid \psi(v(z), w(z)) \in \mathscr{C}(h)\} & (h \in \mathscr{F}_i(0)) \\ \mathscr{D}_i \quad = \{\psi \in \mathscr{D} \mid \psi(v(z), w(z)) \in \mathscr{C}_i\} & (i = 0, 1), \end{cases}$$

where $\mathscr{D}$ denotes the class of tests based on (v, w). The one-sided t-test ϕ_1 in (4.3) is obviously expressed as $\phi_1(z) = \psi_1(v(z))$ with $\psi_1(v) = 1$ if $v > c$ and $\psi_1(v) = 0$ otherwise, so that $\psi_1 \in \mathscr{D}_0 \subset \mathscr{D}_1 \subset \mathscr{D}(h)$. Let $E_0^v(\cdot)$ denote the expectation of $\cdot$ with respect to the density $r_0(v)$ of v in (4.18) where $\mu = 0$, and let a.a. (w, h) denote "almost all w with respect to the density $r_1(w)$ of w" in (4.19) where r_1 depends on h.

Theorem 4.3. *For Problem* [i], *the one-sided t-test ψ_1 is UMP in the class of conditional level α tests*

$$(4.21)\qquad \mathscr{K}(h) = \{\psi \in \mathscr{D}(h) \mid E_0^v[\psi(v, w)] \le \alpha \quad \text{a.a.} \quad (w, h)\}.$$

Proof. Write $h(z) = q(\|z - \mu\underset{\sim}{1}\|^2)$ in $\mathscr{F}_1(\mu)$ and fix $\mu = \mu_1$. Then from (2) of Lemma 4.1, the conditional density of v given w is given by $g(v, w\colon \mu_1)/\int_{-1}^{1} g(v, w\colon \mu_1)\,dv$. By applying the Neyman–Pearson lemma to this conditional density and by using the nonincreasing property of q, the test with critical region $v > c(w)$ is MP in the class (4.21). But from Lemma 4.1 (3), v and w are independent under $\mu = 0$, and so the critical point $c(w)$ can be chosen independently of w. Further, this MP test does not depend on q and μ_1, so it is UMP in the class (4.21), and thus completes the proof.

The class of level α tests in $\mathscr{D}$ in which ψ_1 claims the UMP property is not $\mathscr{D}(h)$ in (4.20) but the subclass $\mathscr{K}(h)$ of conditional level α tests, given w. In fact, no UMP test exists in the class $\mathscr{D}(h)$, because for testing $\mu = 0$ versus $\mu = \mu_1 > 0$ in the density $g(v, w\colon \mu)$ of (4.17), the MP test with critical region

$$g(v, w\colon \mu_1)/g(v, w\colon 0) = q(w - 2\sqrt{w}\,v\sqrt{n}\,\mu_1 + n\mu_1^2)/q(w) > c$$

in general depends on the fixed μ_1 as well as q. On the other hand, the above theorem implies the result in Theorem 4.2 (2). To see this, let

$$(4.22)\qquad \mathscr{K}_i \equiv \bigcap_{h \in \mathscr{F}_i(0)} \mathscr{K}(h) \qquad (i = 0, 1).$$

Lemma 4.2. (1) $\mathscr{D}_0 = \mathscr{K}_0$. (2) $\mathscr{D}_0 = \mathscr{K}_1$.

Proof. (1) If $\psi \in \mathscr{K}(h)$ for all $h \in \mathscr{F}_0(0)$, then $E_0^v[\psi(v, w)] \leq \alpha$ a.a. (w, h) for any $h \in \mathscr{F}_0(0)$, implying $E[\psi(v, w) \mid h] \leq \alpha$ for all $h \in \mathscr{F}_0(0)$. Hence $\psi \in \mathscr{D}_0$. To show the converse, suppose that $\psi \notin \mathscr{K}_0$. Then there exists $h_1 \in \mathscr{F}_0(0)$ such that h_1 gives a positive mass to the set

$$S = \{w > 0 \mid E_0^v[\psi(v, w)] > \alpha\},$$

which itself is independent of any $h \in \mathscr{F}_0(0)$, since the density of v under $\mu = 0$ is independent of h. Let $r_1(w)$ denote the marginal density of w under h_1, which is of the form (4.19). Since $r_1(w)$ is absolutely continuous with respect to the Lebesgue measure, S has a positive Lebesgue measure, which implies that S contains a bounded nonempty open set A. Define a density on R^n by

$$h_2(z) = I_A(\|z\|^2) \Big/ \int_{R^n} I_A(\|z\|^2)\, dz,$$

where $I_A(\cdot)$ denotes the indicator function of A. Then $h_2 \in \mathscr{F}_0(0)$ and h_2 gives the whole mass to the set A. Therefore,

$$E[\psi \mid h_2] = E\{E_0^v[\psi(v, w)] \mid r_2\} > \alpha,$$

where $r_2(w)$ is the density of w under h_2. This implies $\psi \notin \mathscr{D}_0$, completing the proof of (1). Next we prove (2). First $\mathscr{D}_0 \subset \mathscr{K}_1$ follows from (1) and $\mathscr{F}_1(0) \subset \mathscr{F}_0(0)$. Conversely, suppose $\psi \in \mathscr{K}_1$ or, equivalently, $E[\psi \mid h] \leq \alpha$ for all $h \in \mathscr{F}_1(0)$. Then by the completeness of w for the family $\{N(0, \sigma^2 I_n) \mid \sigma^2 > 0\}$ whose densities are included in $\mathscr{F}_1(0)$, $E_0^v[\psi(v, w)] \leq \alpha$ a.e. (Lebesgue), which implies $\psi \in \mathscr{D}_0$. This completes the proof of (2).

Corollary 4.1. (1) *For Problem* [i], ψ_1 *is UMP in* $\mathscr{K}_1 = \mathscr{D}_0$. (2) *For Problem* [ii], ψ_1 *is UMP in* $\mathscr{D}_0$.

Proof. (1) follows from Theorem 4.3 with Lemma 4.2 (2), whereas (2) follows from Theorem 4.3 with Lemma 4.2 (1).

Although Theorem 4.3 implies Corollary 4.1 (2) (or Theorem 4.2 (2)), the latter does not imply the former. In fact, Corollary 4.1 (2) does not imply that for each fixed $h \in \mathscr{F}_1(\mu)$, ψ_1 is UMP in $\mathscr{K}(h)$. In this sense, Theorem 4.3 is stronger than Corollary 4.1 (2).

Next we shall consider Problem [iii]. First note that the difference between Problems [ii] and [iii] lies in the following fact.

Lemma 4.3. *$\mathscr{D}_0$ is a proper subset of $\mathscr{D}_1$, i.e., $\mathscr{D}_0 \subsetneqq \mathscr{D}_1$.*

Proof. Since $\mathscr{F}_0(0) \supset \mathscr{F}_1(0)$, $\mathscr{D}_0 \subset \mathscr{D}_1$ is clear. To show $\mathscr{D}_0 \neq \mathscr{D}_1$, define a test ψ on $(0, \infty)$ by $\psi(w) = 1$ for $1 < w < 1 + a$ $(a > 0)$, and $\psi(w) = 0$ otherwise. Then for $f(z) = q(\|z\|^2)$ in $\mathscr{F}_1(0)$, from (4.19)

$$E_f[\psi(w)] = \int \psi(w) c_0 w^{n/2-1} q(w)\, dw \leq c_0(1 + a)^{n/2-1} q(1) a$$

$$\leq c_0(1 + a)^{n/2-1} a,$$

since $\int q(\|z\|^2)\, dz = 1$, and the nonincreasing property of q implies $q(1) \leq 1$. Hence, for a sufficiently small $a > 0$, $E_f[\psi(w)] \leq \alpha$ for all $f \in \mathscr{F}_1(0)$. On the other hand, $f_1(z) = I_B(\|z\|^2)/\int I_B(\|z\|^2)\, dz$ with $B = \{w \mid 1 < w < 1 + a\}$ belongs to $\mathscr{F}_0(0)$ but not to $\mathscr{F}_1(0)$, and $E_{f_1}[\psi(w)] > \alpha$. This completes the proof.

From this lemma, it follows that Problem [iii] of testing H_1 versus K_1 is in fact different from Problem [ii] of testing H_0 versus K_1. However, it is conjectured that, except for the case $1 \geq \alpha \geq 1/2$, there exists no UMP test for Problem [iii], though ψ_1 is UMP for Problem [ii] for all $0 \leq \alpha \leq 1$. We prove below that ψ_1 is UMP for Problem [iii] when $1 \geq \alpha \geq 1/2$. The following lemma is used in the proof.

Lemma 4.4. *Let $(\mathscr{X}, \mathscr{B}, \nu)$ be a sigma-finite measure space and let $\mathscr{F}$ be a class of densities with respect to ν. For testing $h \in \mathscr{F}$ versus $h = f_1 \notin \mathscr{F}$, a test defined by*

$$\phi_0(x) = \begin{cases} 1 & \text{if } f_1(x) > k f_0(x) \\ \gamma(x) & \quad = \\ 0 & \quad < \end{cases}$$

for some $f_0 \in \mathscr{F}$ is a MP test of level α, provided $E_f[\phi_0] \leq \alpha$ for all $f \in \mathscr{F}$.

Proof. By the Neyman–Pearson lemma, ϕ_0 is MP for testing $h = f_0$ versus $h = f_1$ in the class $\{\phi \in \mathscr{T} \mid E_{f_0}[\phi] \leq \alpha\}$ where $\mathscr{T}$ is the class of tests. But this class contains the class $\{\phi \in \mathscr{T} \mid E_f[\phi] \leq \alpha \text{ for all } f \in \mathscr{F}\}$ to which ϕ_0 belongs. Hence the result follows.

Theorem 4.4. *Assume $\frac{1}{2} \leq \alpha \leq 1$, and fix $f_1(z) = q_1(\|z - \mu \underset{\sim}{1}\|^2)$ in $\mathscr{F}_1(\mu)$ where $\mu > 0$. Then for testing H_1: $h \in \mathscr{F}_1(0)$ versus $h = f_1$ with $\mu > 0$, ψ_1 is UMP in the class $\mathscr{D}_1$.*

Proof. To apply Lemma 4.4, fix $\mu > 0$ and choose as f_0

$$f_0(z) = q_1(\|z\|^2 + 2\beta\mu\sqrt{n}\,\|z\| + n\mu^2)/J,$$

where $0 < \beta < 1$ and $J = \int f_0(z)\,dz$. Since $1 = \int q_1(\|z\|^2)\,dz = c_0 \int_0^\infty r^{n-1} q_1(r^2)\,dr$ from Lemma 4.1,

$$0 < J = c_1 \int_\tau^\infty (r - \tau)^{n-1} q_1(r^2 + \delta)\,dr < \infty,$$

where c_1 is a constant, $\tau = \beta\mu\sqrt{n} > 0$ and $\delta = n\mu^2(1 - \beta^2) > 0$. Further, since $\tau > 0$, $q_1((\|z\| + \tau)^2 + \delta)$ is a nonincreasing function of $\|z\|$. This implies $f_0 \in \mathscr{F}_1(0)$, and hence by Lemma 4.4 with $k = J$ and $\gamma = 0$, the test ϕ_0 with critical region

$$q_1(\|z - \mu\underset{\sim}{1}\|^2) \geq q_1(\|z\|^2 + 2\beta\mu\sqrt{n}\,\|z\| + n\mu^2) \tag{4.23}$$

is MP for testing $\mathrm{H}_1\colon h \in \mathscr{F}_1(0)$ versus $h = f_1$, provided $E_f[\phi_0] \leq \alpha$ for all $f \in \mathscr{F}_1(0)$. Since (4.23) yields the critical region $v \geq -\beta$, and since the distribution of v under $h \in \mathscr{F}_1(0)$ does not depend on h as has been observed in Chapter 3, $v \geq -\beta$ is MP for testing H_1 versus $h = f_1$. But since ϕ_0 is independent of the fixed $\mu > 0$, it is UMP. Noting that $P(v > -\beta) \geq P(v > 0) = 1/2$ for any $h \in \mathscr{F}_1(0)$, the proof is completed.

Corollary 4.2. *For Problem* [iii], ψ_1 *is UMP when* $\frac{1}{2} \leq \alpha \leq 1$.

Proof. This is immediate from Theorem 4.4, since ψ_1 does not depend on the fixed f_1 in $\mathscr{F}_1(\mu)$.

4.3. Two-Sided Testing Problems without Invariance

Again let $z = (z_1, \ldots, z_n)'$ be a random vector with density h, but this time we consider the following two-sided problems:

[iv] For a fixed $h \in \mathscr{F}_2(\mu)$, $\quad \mu = 0 \quad$ versus $\quad \mu \neq 0$,
[v] $\mathrm{H}_2\colon h \in \mathscr{F}_2(0)$ $\quad$ versus $\quad \mathrm{K}_2\colon h \in \mathscr{F}_2(\mu)$, $\quad \mu \neq 0$.

Here $\mathscr{F}_2(\mu)$ is defined in (4.8), and the difference between Problems [iv] and [v] results in the difference between the classes of level α tests under the null hypotheses. For Problem [iv], the class is given by $\mathscr{C}(h)$ in (4.16), whereas for Problem [v], it is given by $\mathscr{C}_2$ in (4.12). As has been done in the case of the one-sided problems in Section 4.2, we shall restrict ourselves to the following classes of tests based on the sufficient statistic

$$(v, w) = (v(z), w(z)) = (z'\underset{\sim}{1}/\|z\|\,\|\underset{\sim}{1}\|, \|z\|^2); \tag{4.24}$$

$$\begin{cases} \mathscr{D}(h) = \{\psi \in \mathscr{D} \mid \psi(v(z), w(z)) \in \mathscr{C}(h)\} & (h \in \mathscr{F}_2(0)) \\ \mathscr{D}_2 \;\; = \{\psi \in \mathscr{D} \mid \psi(v(z), w(z)) \in \mathscr{C}_2\} \end{cases} \tag{4.25}$$

where $\mathscr{D}$ denotes the class of tests based on (v, w). The two sided t-test ϕ_2 in (4.4) is clearly expressed as $\phi_2(z) = \psi_2(v(z))$ with $\psi_2(v) = 1$ if $|v| > c$ and $\psi_2(v) = 0$ otherwise, so that $\psi_2 \in \mathscr{D}_2 \subset \mathscr{D}(h)$. As before, let $E_0^v(\cdot)$ denote the expectation of $\cdot$ with respect to the density $r_0(v)$ of v in (4.18) and let a.a. (w, h) denote "almost all w with respect to the density $r_1(w)$ of w" in (4.19) where r_1 depends on h. Further let

$$k(v\colon \mu, w) = g(v, w\colon \mu) \bigg/ \int g(v, w\colon \mu)\, dv \tag{4.26}$$

be the conditional density of v given w where $g(v, w\colon \mu)$ is given by (4.17). It is noted that $k(v\colon 0, w) = r_0(v)$. Let

$$\pi(\psi, (\mu, w, h)) = E[\psi(v, w) \mid k(\cdot\colon \mu, w)] = \int_{-1}^{1} \psi(v, w) k(v\colon \mu, w)\, dv \tag{4.27}$$

be the conditional power function of ψ, given w, which becomes

$$\pi(\psi, (0, w, h)) = E_0^v[\psi(v, w)] = \int_{-1}^{1} \psi(v, w) r_0(v)\, dv \tag{4.28}$$

under $\mu = 0$, and let $\mathscr{K}\mathscr{U}(h)$ and $\mathscr{K}\mathscr{S}(h)$ be respectively the class of conditional unbiased tests of level α in $\mathscr{D}$ and the class of conditional similar tests of level α in $\mathscr{D}$ where $h \in \mathscr{F}_2(\mu)$:

(4.29)

$$\begin{aligned} \mathscr{K}\mathscr{U}(h) = \{\psi \in \mathscr{D}(h) \mid E_0^v[\psi(v, w)] \le \alpha \qquad & \text{a.a } (w, h), \\ \pi(\psi, (\mu, w, h)) \ge \alpha \qquad & \text{for all} \quad \mu \ne 0 \quad \text{a.a. } (w, h)\} \\ \mathscr{K}\mathscr{S}(h) = \{\psi \in \mathscr{D}_2 \mid E_0^v[\psi(v, w)] = \alpha \qquad & \text{a.a. } (w, h)\}. \end{aligned}$$

Lemma 4.5. *For each fixed* $h \in \mathscr{F}_2(\mu)$, $\mathscr{K}\mathscr{U}(h) \subset \mathscr{K}\mathscr{S}(h)$.

Proof. The conditional power $\pi(\psi, (\mu, w, h))$ is shown to be continuous a.a. (w, h) at $\mu = 0$ (see Exercise 4.1). Hence, $\pi(\psi, (\mu, w, h)) \ge \alpha$ for all $\mu \ne 0$ a. a. (w, h) implies $\pi(\psi, (0, w, h)) = \alpha$. Hence, $\psi \in \mathscr{K}\mathscr{S}(h)$.

Now we shall show that for a fixed h in $\mathscr{F}_2(\mu)$, the two-sided t-test ψ_2 is UMP in a subclass of $\mathscr{K}\mathscr{S}(h)$.

Theorem 4.5. *For Problem* [iv], *the two-sided t-test* ψ_2 *is UMP in the class of tests in* $\mathscr{K}\mathscr{S}(h)$, *satisfying*

$$E_0^v[v\psi(v, w)] = 0 \qquad \text{a.a. } (w, h). \tag{4.30}$$

Proof. Write $h(z) = q(\|z - \mu \underset{\sim}{1}\|^2)$ where q is nonincreasing and convex, and consider the problem of testing $\mu = 0$ versus $\mu = \mu_1 \neq 0$ in the conditional density $k(v\colon \mu, w)$ in (4.26). Then $\mathscr{K}\mathscr{S}(h)$ is nothing but the class of *size* α tests for this problem. Hence by the generalized Neyman–Pearson lemma (see Chapter 2), a test ψ^*, which maximizes the conditional power $\pi(\psi, (\mu_1, w, h))$ subject to (4.30), is given by

$$\psi^*(v, w) = \begin{cases} 1 & \text{if } k(v\colon \mu, w) > c_1 r_0(v) + c_2 v r_0(v) \\ 0 & \text{otherwise,} \end{cases} \tag{4.31}$$

where c_1 and c_2 are chosen to satisfy $E_0^v(\psi^*) = \alpha$ and (4.30). Now noting from the convexity of q that

$$[k(v\colon \mu_1, w)/r_0(v)] - c_2 v$$

is convex in v, the test ψ^* in (4.31) is given by

$$\psi^*(v, w) = \begin{cases} 1 & \text{if} \quad v > b \quad \text{or} \quad v < a \\ 0 & \text{if} \quad a \leq v \leq b, \end{cases} \tag{4.32}$$

where a and b may depend on w and are chosen to satisfy $E_0^v(\psi^*) = \alpha$ and (4.30). It is clear that $\psi^*(v, w)$ is independent of μ_1. Moreover, a and b satisfying these conditions for any w must be independent of w and satisfy $-a = b = c$ (Exercise 4.2). Therefore, $\psi^* = \psi_2$, and hence for any $\psi \in \mathscr{K}\mathscr{S}(h)$ satisfying (4.30) and for any $\mu \neq 0$,

$$\pi(\psi_2, (\mu, w, h)) \geq \pi(\psi, (\mu, w, h)) \quad \text{a.a. } (w, h).$$

Taking expectation from both sides of this inequality proves the theorem.

Corollary 4.3. *For Problem* [iv], ψ_2 *is UMP in the class of tests in* $\mathscr{K}\mathscr{U}(h)$ *satisfying* (4.30).

Proof. Immediate from $\psi_2 \in \mathscr{K}\mathscr{U}(h)$.

Next, to consider Problem [v], let

$$\mathscr{K}\mathscr{U} = \bigcap_{h \in \mathscr{F}_2(0)} \mathscr{K}\mathscr{U}(h) \qquad \text{and} \quad \mathscr{K}\mathscr{S} = \bigcap_{h \in \mathscr{F}_2(0)} \mathscr{K}\mathscr{S}(h).$$

Also let $\mathscr{U}$ and $\mathscr{S}$, respectively, be the class of unbiased tests of level α in $\mathscr{D}_2$ and the class of similar tests of level α in $\mathscr{D}_2$:

$$\begin{aligned} \mathscr{U} = \{\psi \in \mathscr{D}_2 \mid \pi(\psi, (0, h)) \le \alpha \quad & \text{for any} \quad h \in \mathscr{F}_2(0) \\ \pi(\psi, (\mu, h)) \le \alpha \quad & \text{for any} \quad h \in \mathscr{F}_2(\mu) \quad \text{and} \quad \mu \neq 0\} \\ \mathscr{S} = \{\psi \in \mathscr{D}_2 \mid \pi(\psi, (0, h)) = \alpha \quad & \text{for any} \quad h \in \mathscr{F}_2(0)\}, \end{aligned}$$

where $\pi(\psi, (\mu, h)) = E[\psi(v, w)]$ is the power function of ψ under $h \in \mathscr{F}_2(\mu)$.

Lemma 4.6. (1) $\mathscr{K}\mathscr{U} \subset \mathscr{U}$, (2) $\mathscr{U} \subset \mathscr{S}$, *and* (3) $\mathscr{K}\mathscr{S} = \mathscr{S}$.

Proof. (1) $\mathscr{K}\mathscr{U} \subset \mathscr{U}$ follows from the definitions of $\mathscr{K}\mathscr{U}$ and $\mathscr{U}$. (2) follows from the fact: $\psi \in \mathscr{U}$ implies $\pi(\psi, (0, h)) = \alpha$ for all $h \in \mathscr{F}_2(0)$ from the continuity of $\pi(\psi, (\mu, h))$ in μ. To show (3), note that $\mathscr{N}(\mu) = \{N(\mu, \sigma^2) \mid \sigma^2 > 0\} \subset \mathscr{F}_2(\mu)$ and that $w = \|z\|^2$ is complete for $\mathscr{N}(0) \subset \mathscr{F}_2(0)$. Hence $\pi(\psi, (0, h)) = E^w[E_0^v[\psi \mid w]] = \alpha$ implies

$$E_0^v[\psi(v, w)] = \alpha \qquad \text{a.a. } (w, h) \qquad \text{for all} \quad h \in \mathscr{F}_2(0),$$

which shows $\mathscr{S} \subset \mathscr{K}\mathscr{S}$. On the other hand, $\mathscr{S} \supset \mathscr{K}\mathscr{S}$ follows from the definition of $\mathscr{K}\mathscr{S}$. This completes the proof.

Let $\mathscr{U}^*$ and $\mathscr{S}^*$ be respectively the class of tests in $\mathscr{U}$ satisfying (4.30) and the class of tests in $\mathscr{S}$ satisfying (4.30).

Theorem 4.6. *For Problem* [v], *the two-sided t-test* ψ_2 *is UMP in* $\mathscr{S}^*$, *and hence in* $\mathscr{U}^*$.

Proof. By Theorem 4.5, for each $h \in \mathscr{F}_2(\mu)$, ψ_2 is UMP in the class of tests in $\mathscr{K}\mathscr{S}(h)$, satisfying (4.30). Hence, for any $h \in \mathscr{F}_2(\mu)$, ψ_2 is UMP in the class of tests in $\mathscr{S} = \mathscr{K}\mathscr{S}$ satisfying (4.30). Because $\psi_2 \in \mathscr{U} \subset \mathscr{S}$, this completes the proof.

As will be shown in the next section, when the invariance principle is applied to Problems [iv] and [v], the class $\mathscr{I}$ of invariant tests of size α in $\mathscr{D}$ is the class of tests of size α based on v only, and ψ_2 is UMPI (see also Section 3.5 of Chapter 3). Hence if $\mathscr{U}^* = \mathscr{I}$, the result in Theorem 4.6 is not stronger than the UMPI property of ψ_2. However $\mathscr{U}^*$ contains a test which depends nontrivially on w, and so $\mathscr{U}^* \supsetneqq \mathscr{I}$.

4.4. UMPI Property of *t*-Test

As has been shown in Chapter 3, when the density of z is given by

$$h(z \mid \mu, \sigma^2) = \sigma^{-n} q(\|z - \mu \underset{\sim}{1}\|^2/\sigma^2), \tag{4.33}$$

the one sided t-test ϕ_1 in (4.3) is UMPI for the one-sided problem, $\mu = 0$ versus $\mu > 0$, and the two-sided t-test ϕ_2 in (4.4) is UMPI for the two-sided problem, $\mu = 0$ versus $\mu \neq 0$, where q is unknown but fixed. The groups leaving the problems invariant are respectively given by $\mathscr{G}_1 \equiv R_+ \times \tilde{\mathcal{O}}(n)$ and $\mathscr{G}_2 \equiv R_* \times \tilde{\mathcal{O}}(n)$, where $R_+ = \{a > 0\}$, $R_* = \{a \in R \mid a \neq 0\}$,

$$\tilde{\mathcal{O}}(n) = \{\Gamma \in \mathcal{O}(n) \mid \Gamma \underset{\sim}{1} = \underset{\sim}{1}\},$$

and $\mathscr{G}_i$ acts on z by $(\gamma, \Gamma)(z) = \gamma \Gamma z$ for $(\gamma, \Gamma) \in \mathscr{G}_i$ $(i = 1, 2)$. Then maximal invariants under $\mathscr{G}_1$ and $\mathscr{G}_2$ are, respectively, v and $|v|$ with v in (4.2). Therefore the classes of invariant tests of level α are respectively given by

$$\mathscr{I}_1 = \{\psi \in \mathscr{D} \mid \psi \quad \text{is based on} \quad v \quad \text{only}, \qquad E_0^v[\psi] \leq \alpha\} \tag{4.34}$$

and

$$\mathscr{I}_2 = \{\psi \in \mathscr{I}_1 \mid \psi(-v) = \psi(v)\}. \tag{4.35}$$

Since the null distribution of v does not depend on h in (4.33) (Lemma 4.1), it follows from (4.20) and (4.25) that

$$\mathscr{I}_2 \subset \mathscr{I}_1 \subset \mathscr{D}_0 \subset \mathscr{D}_1 \subset \mathscr{D}(h). \tag{4.36}$$

In fact, the following lemma holds.

Lemma 4.7. $\mathscr{I}_1 \subsetneqq \mathscr{D}_0$.

Proof. Since $\mathscr{I}_1 \subset \mathscr{D}_0$, we show $\mathscr{I}_1 \neq \mathscr{D}_0$. Take any $\psi_1(w) \not\equiv$ const., such that $0 \leq \psi_1(w) \leq 1$, and take any $\psi_2(v)$ in $\mathscr{I}_1$. Then the test defined by

$$\psi(w, v) = \psi_1(w)\psi_2(v)$$

belongs to $\mathscr{D}_0$ because $E_0^v[\psi(w, v)] = \psi_1(w)E_0^v[\psi_2(v)] \leq \alpha$. But $\psi \notin \mathscr{I}_1$, completing the proof.

Consequently the statement in Theorem 4.3 that ψ_1 is UMP in the class $\mathscr{K}(h)$ of conditional level α tests is stronger than the statement that ψ_1 is UMPI because $\mathscr{D}_0 \subset \mathscr{K}(h)$. Therefore the results in Corollary 4.1 are stronger than the result that ψ_1 is UMPI.

On the other hand, for the two-sided problem, it has been shown in Theorem 4.5 that ψ_2 is UMP in $\mathscr{K}\mathscr{S}(h)$ satisfying (4.30) and in Theorem 4.6 that ψ_2 is UMP in $\mathscr{U}^*$. The next lemma implies that the result in Theorem 4.6 is stronger than the result that ψ_2 is UMPI. In fact, $\mathscr{U}^* \subset \mathscr{S}^*$, and $\mathscr{S}^* \cap \mathscr{I}_2$ forms a complete class in $\mathscr{I}_2$.

Lemma 4.8. $\mathscr{S}^* \cap \mathscr{I}_2 \subsetneq \mathscr{S}^*$.

Proof. Note that for $h \in \mathscr{F}_2(0)$, v and w are independent and the densities are given by $r_0(v)$ and $r_1(w)$, respectively, and that the conditions $E_0^v[\psi] = \alpha$ and (4.30) are respectively given by

$$\int \psi(v, w) r_0(v)\, dv = \alpha \qquad \text{a.a. } (w, h \in \mathscr{F}_2(0)) \tag{4.37}$$

and

$$\int v\psi(v, w) r_0(v)\, dv = 0 \qquad \text{a.a. } (w, h \in \mathscr{F}_2(0)). \tag{4.38}$$

Now take two tests, $\rho_1(v)$ and $\rho_2(v)$, based on v only, which are functionally independent and satisfy (4.38) with

$$\beta_i = E_0^v[\rho_i(v)] = \int \rho_i(v) r_0(v)\, dv \qquad (i = 1, 2),$$

where $\beta_1 > \alpha > \beta_2 > 0$. For such $\rho_i(v)$ then choose any two tests, $\gamma_1(w)$ and $\gamma_2(w)$, based on w only, which satisfy

$$\beta_1\gamma_1(w) + \beta_2\gamma_2(w) \equiv \alpha.$$

With this choice of $\rho_i(v)$'s and $\gamma_i(w)$'s, finally define a test based on (v, w) by

$$\phi(v, w) = \rho_1(v)\gamma_1(w) + \rho_2(v)\gamma_2(w).$$

Then ϕ satisfies $E_0^v[\phi(v, w)] = \alpha$ and ϕ ($\not\equiv$ const.) does depend nontrivially on w. Therefore, $\phi \in \mathscr{S}^* - \mathscr{I}_2$, completing the proof.

4.5. Tests on Serial Correlation without Invariance

In this section, proceeding in a manner similar to that in the case of the t-tests, we deal with the problem of tests on serial correlation without invariance. This problem was dealt with in Example 3.6 of Chapter 3 from the point of view of invariance. The notation used in this section is independent of that in the previous sections. Let

$$(4.39)\qquad \mathcal{Q} = \left\{ q \,\middle|\, q\colon [0,\infty) \to [0,\infty), \qquad \int q(\|x\|^2)\,dx = 1 \right\},$$

and for $\Sigma \in \mathcal{S}(n)$, define the three classes of densities on R^n:

$$(4.40)\qquad \mathcal{F}_0(\Sigma) = \{ f \mid f(x) = |\Sigma|^{-1/2} q(x'\Sigma^{-1}x), \qquad q \in \mathcal{Q} \}$$

$$(4.41)\qquad \mathcal{F}_1(\Sigma) = \{ f \mid f(x) = |\Sigma|^{-1/2} q(x'\Sigma^{-1}x), \qquad q \in \mathcal{Q}, \quad q \text{ is nonincreasing} \}$$

and

$$(4.42)\qquad \mathcal{F}_2(\Sigma) = \{ f \mid f(x) = |\Sigma|^{-1/2} q(x'\Sigma^{-1}x), \qquad q \in \mathcal{Q}, \quad q \text{ is nonincreasing and convex} \}.$$

Clearly $\mathcal{F}_2(\Sigma) \subset \mathcal{F}_1(\Sigma) \subset \mathcal{F}_0(\Sigma)$ for each $\Sigma \in \mathcal{S}(n)$, and $\mathcal{F}_i(a\Sigma) = \mathcal{F}_i(\Sigma)$, $i = 0, 1, 2$ for any $a > 0$. Let z be an $n \times 1$ random vector with density h and consider the testing problems

$$(4.43)\qquad \mathrm{H}_i\colon h \in \mathcal{F}_i(I) \qquad \text{versus} \qquad h \in \mathcal{F}_i(\Sigma_0) \quad (i = 0, 1, 2),$$

where $\Sigma_0 \in \mathcal{S}(n)$ is fixed, and $\Sigma_0 \neq aI$ for any $a > 0$. Let

$$(4.44)\qquad \mathcal{C}_i = \{ \phi \in \mathcal{C} \mid E_f[\phi] \leq \alpha \qquad \text{for all} \quad f \in \mathcal{F}_i(I) \}$$

be the class of level α tests under H_i $(i = 0, 1, 2)$ where $\mathcal{C}$ is the set of all tests on R^n, $0 < \alpha < 1$, and $E_f[\phi] \equiv E[\phi \mid f]$ denotes the expectation of ϕ under f. It is noted that $\mathcal{F}_2(I) \subset \mathcal{F}_1(I) \subset \mathcal{F}_0(I)$ implies $\mathcal{C}_0 \subset \mathcal{C}_1 \subset \mathcal{C}_2$. As in the case of the Student's t-problem, we consider the present problem in terms of the sufficient statistic

$$(4.45)\qquad (t, w) = (z'\Sigma^{-1}z/z'z, \|z\|^2)$$

for $\mathcal{F}_i(I) \cup \mathcal{F}_i(\Sigma)$. Hence let $\mathcal{D}_i$ be the class of the level α tests based on (t, w) in $\mathcal{C}_i$, $i = 0, 1, 2$ so that $\mathcal{D}_0 \subset \mathcal{D}_1 \subset \mathcal{D}_2$. Further let $\lambda_1 \geq \cdots \geq \lambda_n$ be the latent roots of Σ_0 and let $\lambda = (\lambda_1, \ldots, \lambda_n)'$. Note that $\lambda_1 \neq \lambda_n$.

Lemma 4.9. *Let $h(z) = |\Sigma|^{-1/2} q(z'\Sigma^{-1}z)$ be a density in $\mathcal{F}_i(\Sigma)$, $i = 0, 1, 2$.*

(1) *The joint density of* (t, w) *is given by*

$$g(t, w\colon \lambda) = c_0\left(\prod_{i=1}^{n} \lambda_i\right)^{-1/2} q(wt)r_0(t)w^{n/2-1}, \tag{4.46}$$

where $c_0 = \Gamma(\frac{1}{2})^n/\Gamma(\frac{n}{2})$ *and* $r_0(t)$ *is the density of* $\sum \lambda_i T_i$ *on* $[\lambda_n, \lambda_1]$ *with* $(T_1, \ldots, T_n)$ *obeying the Dirichelet distribution* $D_n(\frac{1}{2}, \ldots, \frac{1}{2}, \frac{1}{2})$ *(see Chapter 1).* (2) *Under* H_i, t *and* w *are independent and their densities are given by* $r_0(t)$ *and* $c_0 q(w)w^{n/2-1}$ $(i = 0, 1, 2)$, *respectively.*

Proof. (1) We proceed in a similar way as in the proof of Lemma 4.1. By taking $P \in \mathcal{O}(n)$ such that $P'\Sigma P = \mathrm{diag}\{\lambda_{n-1}, \lambda_{n-2}, \ldots, \lambda_2, \lambda_n, \lambda_1\}$, changing z into $y = Pz$, and using the polar coordinates $\{s_i(\theta)\}$ given in the proof of Lemma 4.1 with $y_i = \sqrt{w}\, s_i(\theta)$, the density of $(w, \theta_1, \ldots, \theta_{n-1})$ is given by

$$C(n, \lambda)q(wt(\theta))w^{n/2-1}J(\theta)$$

where $C(n, \lambda)$ is a constant, $J(\theta)$ is the Jacobian of the transformation, and

$$t(\theta) = \lambda_{n-1}s_1(\theta)^2 + \lambda_{n-2}s_2(\theta)^2 + \cdots + \lambda_2 s_{n-2}(\theta)^2 + \lambda_n s_{n-1}(\theta)^2 + \lambda_1 s_n(\theta)^2.$$

Then transforming $(w, \theta_1, \ldots, \theta_{n-1})$ into $(w, \theta_1, \ldots, \theta_{n-2}, t)$ for each region of $\theta_{n-1} \in (-\pi, -\pi/2)$, $\theta_{n-1} \in [-\pi/2, 0)$, $\theta_{n-1} \in [0, \pi/2)$ and $\theta_{n-1} \in [\pi/2, \pi)$ yields the density in the form $C(n, \lambda)q(wt)w^{n/2-1}R(t, \theta_1, \ldots, \theta_{n-2})$. Therefore, the density of (t, w) is given by (4.46) with $r_0(t) = \int\cdots\int R(t, \theta_1, \ldots, \theta_{n-2}) \times d\theta_1 \ldots d\theta_{n-1}$. The density $r_0(t)$ is identified with that of $\sum \lambda_i T_i$ by considering the case $\Sigma = I$ (see Chapter 1). The proof of (2) follows from that of (1).

Now in order to state our first result, let

$$\mathscr{C}(h) = \{\psi \in \mathscr{D} \mid E_h[\psi \mid \Sigma = I] \le \alpha\} \tag{4.47}$$

be the class of level α tests under a density h with $\Sigma = I$, where $\mathscr{D}$ denotes the class of tests based on (t, w), and define

$$\psi_0(t) = \begin{cases} 1 & \text{if} \quad t < c \\ 0 & \text{if} \quad t \ge c, \end{cases} \tag{4.48}$$

with $E_0^t(\psi_0) = \alpha$, so that $\psi_0 \in \mathscr{D}_0 \subset \mathscr{D}_1 \subset \mathscr{D}_2 \subset \mathscr{D}(h)$, where $E_0^t(\cdot)$ denotes the expectation of $\cdot$ with respect to the density $r_0(t)$ of t in (4.46).

Theorem 4.7. (1) *For a fixed* $h \in \mathscr{F}_1(\Sigma)$, *the test* ψ_0 *in* (4.48) *is MP for testing* $\Sigma = I$ *versus* $\Sigma = \Sigma_0$ *in the class of conditional level* α *tests*

$$\mathscr{K}(h) = \{\psi \in \mathscr{D}(h) \mid E_0^t[\psi(t, w)] \le \alpha \qquad \text{a.a. } (w, h)\}. \tag{4.49}$$

(2) *For testing* $\mathrm{H}_1: h \in \mathscr{F}_1(I)$ *versus* $\mathrm{K}_1: h \in \mathscr{F}_1(\Sigma_0)$, *the test* ψ_0 *in* (4.48) *is UMP in the class* $\mathscr{K}_1 \equiv \bigcap_{h \in \mathscr{F}_1(I)} \mathscr{K}(h)$.

Proof. (1) By (1) of Lemma 4.9, the conditional density of t given w is given by

$$k(t: w, \lambda) = g(t, w: \lambda) \Big/ \int_{\lambda_n}^{\lambda_1} g(t, w: \lambda)\, dt. \tag{4.50}$$

Hence, by applying the Neyman–Pearson lemma to this density and using the nonincreasing property of q, we obtain the MP test with the critical region $t > c(w)$. But under $\Sigma = I$, t and w are independent and hence the cut-off point can be chosen to be independent of w. Therefore (1) follows. For (2), it suffices to observe that the MP test does not depend on the fixed h and that the class of tests in (2) becomes $\mathscr{K}_1$.

Theorem 4.7 (1) is similar to Theorem 4.3, whereas the next lemma is quite similar to Lemma 4.2 (Exercise 4.3).

Lemma 4.10. $\mathscr{D}_0 = \mathscr{K}_0 = \mathscr{K}_1$ *where* $\mathscr{K}_i = \bigcap_{h \in \mathscr{F}_i(0)} \mathscr{K}(h)$ $(i = 0, 1)$.

Consequently, ψ_0 is UMP for the problem of testing H_0 versus K_1 in the class of level α tests $\mathscr{D}_0$. However, ψ_0 is not UMP for the problem of testing H_1 versus K_1 in the class of level α tests $\mathscr{D}_1$, though it is UMP in $\mathscr{D}_0 = \mathscr{K}_1$ for the same problem. In fact, by proceeding as we did in Lemma 4.3, we obtain the following result (Exercise 4.4)

Lemma 4.11. $\mathscr{D}_0 = \mathscr{K}_1 \subsetneq \mathscr{D}_1$.

In the above argument, a simple alternative is considered for Σ. However, in applications, Σ is unknown, though it may be a function of some parameters. In particular, in a heteroscedastic model and in a serial correlation model, Σ is specified as

$$\Sigma(\delta) = \gamma\Phi(\delta) \qquad \text{with} \quad \Phi(\delta)^{-1} = I_n + \delta A, \tag{4.51}$$

where $\delta \in \Lambda \equiv \{\delta \in R \mid \Sigma(\delta)^{-1} \in \mathscr{S}(n)\}$. Here A is an $n \times n$ known matrix. When $z \sim N(0, \gamma\Phi(\delta))$, it is shown in Anderson (1948) that for testing the hypothesis $\delta = 0$ versus the alternative $\delta > 0$, the test which rejects the null hypothesis for small values of

$$v = z'Az/z'z \tag{4.52}$$

is UMP, whereas for testing $\delta = 0$ versus $\delta \neq 0$, the test with critical region $v < c_1$ or $v > c_2$ is UMPU. Here we shall consider robustness properties of these tests without invariance in the above framework.

Clearly, (v, w) forms a sufficient statistic where v is defined in (4.52) and $w = \|z\|^2$ as before. Let $d_1 \geq \cdots \geq d_n$ be the latent roots of A with $d_1 \neq d_n$. Then the following result on the joint distribution of (v, w) is basic to our subsequent discussion. Its proof, which is similar to that of Lemma 4.9, is left as an exercise (Exercise 4.5).

Lemma 4.12. *Let $h(z) = \gamma^{-n/2}|I + \delta A|^{1/2} q([z'z + \delta z'Az]/\gamma)$ be a density in $\mathscr{F}_0(\gamma\Phi(\delta))$. Then the joint density of (v, w) is given by*

$$(4.53) \qquad \tilde{g}(v, w: \gamma, \delta) = c_0 \gamma^{-n/2}|I + \delta A|^{1/2} q([w + \delta v w]/\gamma) w^{n/2-1} \tilde{r}_0(v),$$

where $c_0 = \Gamma(\frac{1}{2})^n/\Gamma(\frac{n}{2})$, and $\tilde{r}_0(v)$ is the density of $\sum d_i T_i$ on $[d_n, d_1]$, with $(T_1, \ldots, T_n)$ obeying the Dirichelet distribution $D_n(\frac{1}{2}, \ldots, \frac{1}{2}, \frac{1}{2})$. Further, under $\delta = 0$, v and w are independent and their densities are given by $\tilde{r}_0(v)$ and $c_0 q(w/\gamma) w^{n/2-1}$, respectively.

Now let

$$\tilde{\mathscr{D}}(h) = \{\psi \in \tilde{\mathscr{D}} \mid \pi(\psi, (\gamma, 0, h)) \leq \alpha \qquad \text{for all} \quad \gamma > 0\}$$

be the class of level α tests under h in $\mathscr{F}_0(\gamma\Phi(\delta))$ with $\delta = 0$, where $\tilde{\mathscr{D}}$ denotes the class of tests based on (v, w) and $\pi(\psi, (\gamma, \delta, h)) = \iint \psi \tilde{g}(v, w: \gamma, \delta)\, dv\, dw$. Let $E_0^v(\cdot)$ denote the expectation of $\cdot$ with respect to the density $\tilde{r}_0(v)$ in Lemma 4.12.

Theorem 4.8. (1) *For a fixed $h \in \mathscr{F}_1(\gamma\Phi(\delta))$, the test with critical region $v < c$ is UMP for testing $\delta = 0$ versus $\delta > 0$ in the class of conditional level α tests,*

$$(4.54) \qquad \tilde{\mathscr{K}}(h) = \{\psi \in \tilde{\mathscr{D}}(h) \mid E_0^v[\psi(v, w)] \leq \alpha \qquad \text{a.a. } (w, h)\}.$$

(2) *For a fixed $h \in \mathscr{F}_2(\gamma\Phi(\delta))$, the test with critical region $v < c_1$ or $v > c_2$ is UMP for testing $\delta = 0$ versus $\delta \neq 0$ in the class of tests satisfying*

$$(4.55) \qquad \begin{cases} E_0^v[\psi(v, w)] = \alpha & \text{a.a } (w, h) \\ E_0^v[v\psi(v, w)] = \alpha \operatorname{tr} A/n & \text{a.a. } (w, h). \end{cases}$$

Proof. (1) Similar to the proof of Theorem 4.3. (2) We proceed in a manner similar to that in the proof of Theorem 4.5. Let

$$\tilde{k}(v: w, \gamma, \delta) = \tilde{g}(v, w: \gamma, \delta) \bigg/ \int_{d_n}^{d_1} \tilde{g}(v, w: \gamma, \delta)\, dv$$

be the conditional density of v, given w, where $\tilde{g}$ is given by (4.53) with q convex. With this conditional density, consider now the problem of testing $\delta = 0$ versus $\delta = \delta^*(\neq 0)$, where γ is arbitrarily fixed. Noting that

$\tilde{k}(v; w, \gamma, 0) = \tilde{r}_0(v)$ and by applying the generalized Neyman–Pearson lemma, an MP test ϕ_1 is of the form:

$$\phi_1 = 1 \qquad \text{if} \quad \tilde{k}(v\colon w, \gamma, \delta^*) > a_1\tilde{r}_0(v) + a_2 v\tilde{r}_0(v)$$

and $\phi_1 = 0$ otherwise, where a_1 and a_2 are chosen so that ϕ_1 satisfies (4.55). Since q is convex, $\phi_1 = 1$ either if 1) $v < c$ or $v > c'$, if 2) $v < c$, or if 3) $v > c'$. Here c and c' can be free from the fixed w, since $\tilde{k}(v\colon w, \gamma, 0) = \tilde{r}_0(v)$. We shall show that the last two critical regions 2) and 3) cannot satisfy (4.55). For example, suppose the critical region 2) $v < c$ satisfies it. Then from $E_0^v(v) = \operatorname{tr} A/n$, the second equation in (4.55) is $\int_{d_n}^c v\tilde{r}_0(v)\,dv = R(c)E_0^v(v)$ where $R(c) = \int_{d_n}^c \tilde{r}_0(v)\,dv$. Since

$$\int_{d_n}^c v\tilde{r}_0(v)\,dv = cR(c) - \int_{d_n}^c R(t)\,dt,$$

$$E_0^v(v) = c - [R(c)]^{-1}\int_{d_n}^c R(t)\,dt. \tag{4.56}$$

Let $H(c)$ be the right side of (4.56). Then $H(c)$ is a strictly increasing and continuous function of c with $H'(c) > 0$ where $d_n \le c \le d_1$. Hence because $H(d_1) = E_0^v(v)$, (4.56) is impossible unless $c = d_1$. But $c = d_1$ contradicts $\alpha < 1$. Similarly, the critical region 3) $v > c'$ cannot satisfy (4.55). Therefore, the test with critical region 1) $v < c$ or $v > c'$ is MP and, since it does not depend on δ^* and γ, it is UMP. This completes the proof.

The class of tests satisfying (4.55), which we shall call $\mathscr{K}\mathscr{S}^*(h)$ below, is the class of conditional similar tests of level α satisfying the second equation of (4.55). This is implied by the completeness of (v, w) when h is a normal density.

Next we consider the problems:

$$\text{[A]} \quad \mathrm{H}_1\colon h \in \mathscr{F}_1(\gamma I), \quad \gamma > 0 \quad \text{versus} \quad h \in \mathscr{F}_1(\gamma\Phi(\delta)), \quad \gamma > 0, \quad \delta > 0,$$

$$\text{[B]} \quad \mathrm{H}_2\colon h \in \mathscr{F}_2(\gamma I), \quad \gamma > 0 \quad \text{versus} \quad h \in \mathscr{F}_2(\gamma\Phi(\delta)), \quad \gamma > 0, \quad \delta \neq 0.$$

Since $\mathscr{F}_i(aI) = \mathscr{F}_i(I)$ for all $a > 0$, the class of level α tests under H_i is given by $\mathscr{C}_i$ mentioned in (4.44) where $i = 0, 1, 2$ with $\mathrm{H}_0\colon h \in \mathscr{F}_0(\gamma I)$, $\gamma > 0$. Let

$$\tilde{\mathscr{K}}_1 = \bigcap_{h \in \mathscr{F}_1(I)} \mathscr{K}(h) \qquad \text{and} \qquad \widetilde{\mathscr{K}\mathscr{S}}^* = \bigcap_{h \in \mathscr{F}_2(I)} \mathscr{K}\mathscr{S}^*(h),$$

and let $\tilde{\mathscr{S}}$ be the class of similar tests of level α in $\mathscr{C}_2$. Then proceeding as in the proofs of Lemmas 4.2 and 4.6, we obtain (Exercise 4.6)

Lemma 4.13. (1) $\mathscr{C}_0 = \tilde{\mathscr{K}}_1$ *and* (2) $\widetilde{\mathscr{K}\mathscr{S}}^* = \mathscr{S}^*$.

Therefore, as a corollary to Theorem 4.8, the following theorem follows.

Theorem 4.9. (1) *For Problem* [*A*], *the test* ϕ_A *with critical region* $v < c$ *is UMP in* $\mathscr{C}_0 = \tilde{\mathscr{K}}_1$.
(2) *For Problem* [*B*], *the test* ϕ_B *with critical region* $v < c_1$ *or* $v > c_2$ *is UMP in* $\tilde{\mathscr{K}}\mathscr{S}^*$.

We remark that ϕ_A is not UMP in its relevant class of level α tests $\mathscr{C}_1$ which includes $\mathscr{C}_0$ (see Section 4.2).

Finally let us consider conditions under which Theorems 4.8 and 4.9 still hold for a reparametrization of γ and δ:

$$\tau(\theta_1, \theta_2) = (\gamma(\theta_1, \theta_2), \delta(\theta_1, \theta_2)),$$

where $(\theta_1, \theta_2) \in \Theta_1 \times \Theta_2$ and the Θ_i's are open intervals of R^1. Suppose τ satisfies the conditions: (1) τ is a continuous function from $\Theta_1 \times \Theta_2$ into R^2 such that the image $\tau(\Theta_1 \times \Theta_2)$ contains an open set in $(0, \infty) \times \Lambda$ and $\gamma(\theta_1, \theta_2) > 0$ for all $(\theta_1, \theta_2) \in \Theta_1 \times \Theta_2$; (2) there exists a unique point $\theta_2^* \in \Theta_2$ such that $\lambda(\theta_1, \theta_2^*) = 0$ for all $\theta_1 \in \Theta_1$ and such that $\tau(\theta_1, \theta_2^*)$ is an interior point of the image for each θ_1; and (3) $\Phi(\delta(\theta_1, \theta_2))^{-1} = I + \delta(\theta_1, \theta_2)A > 0$ for all $(\theta_1, \theta_2) \in \Theta_1 \times \Theta_2$. Then it is not difficult to see that the above results hold in terms of (θ_1, θ_2).

Example 4.1. *Circular serial correlation.* Let $z = (z_1, \ldots, z_n)'$ be generated by $z_j = \rho z_{j-1} + u_j$ with $z_0 = z_n$ $(j = 1, \ldots, n)$. Here the density of $u = (u_1, \ldots, u_n)'$ is assumed to belong to $\mathscr{F}_i(\sigma^2 I)$ $(i = 1$ or $2)$, which contains $N(0, \sigma^2 I)$. Since $z = Bz + u$ with

$$B = \begin{bmatrix} 0 & \cdots & \cdots & 1 \\ 1 & \ddots & & \vdots \\ \vdots & \ddots & \ddots & \vdots \\ 0 & \cdots & 1 & 0 \end{bmatrix},$$

the density of z, say h, belongs to $\mathscr{F}_i(\gamma\Phi(\delta))$, where $i = 1, 2$,

$$\gamma = \gamma(\sigma^2, \rho) = \sigma^2/(1 + \rho^2), \qquad \delta = \delta(\sigma^2, \rho) = \rho/(1 + \rho^2)$$

and $\Phi(\delta)^{-1} = I + \delta A$ with $A = -(B + B')$. Hence by Theorem 4.8, for each $h \in \mathscr{F}_1(\gamma\Phi(\delta))$, the test ϕ_A with critical region $v < c$ is UMP for testing $\rho = 0$ versus $\rho > 0$ in the class of conditional level α tests in (4.54), whereas for each $h \in \mathscr{F}_2(r\Phi(\delta))$, the test ϕ_B with critical region $v < c_1$ or $v > c_2$ is UMP for testing $\rho = 0$ versus $\rho \neq 0$ in the class of tests satisfying (4.55). By Theorem

4.9, for testing $h \in \mathscr{F}_1(\gamma I)$, $\gamma > 0$ versus $h \in \mathscr{F}_1(\gamma\Phi(\delta))$, $\gamma > 0$, $\delta > 0$, ϕ_A is UMP in $\mathscr{C}_0 = \tilde{\mathscr{K}}_1$, and for testing $h \in \mathscr{F}_2(\gamma I)$, $\gamma > 0$ versus $h \in \mathscr{F}_2(\gamma\Phi(\delta))$, $\gamma > 0$, $\delta \neq 0$, ϕ_B is UMP similar.

Exercises

4.1. Show that the conditional power $\pi(\psi, (\mu, w, h))$ defined in (4.27) is continuous a.a. (w, h) at $\mu = 0$.

4.2. Show that if $\psi^*(v, w)$ in (4.32) is subject to $E_0^v(\psi^*) = \alpha$ and $E_0^v(v\psi^*) = 0$ a.a. (w, h), then $-a = b$.

4.3. Prove Lemma 4.10. [Hint: Similar to Lemma 4.2.]

4.4. Prove Lemma 4.11. [Hint: Similar to Lemma 4.3.]

4.5. Prove Lemma 4.12. [Hint: Similar to Lemma 4.9.]

4.6. Prove Lemma 4.13. [Hint: Similar to Lemma 4.2 and Lemma 4.6.]

Chapter 5 General Multivariate Analysis of Variance (GMANOVA)

5.1. Introduction

The general multivariate analysis of variance model or, briefly, the GMANOVA model, was originally formulated as a general model for describing the growth curves of animals, and it is often called a growth-curve model. However, because this model includes the classic multivariate analysis of variance (MANOVA) model and various multivariate regression models, it has a broader application in many fields. For example, as mentioned in Section 3.2 of Chapter 3, the MANOVA problem alone covers many interesting applications, such as problems of testing on means, regression coefficients, effects of treatments, etc., in univariate or multivariate models. However, there are many other important testing problems not in the framework of the MANOVA problem that are included in the GMANOVA problem (see Section 3.2A, Chapter 3 for a description of these problems). In addition, from a theoretical point of view, the GMANOVA problem possesses some special features that the MANOVA problem does not; consequently, the GMANOVA problem requires special attention. This point is explained in Sections 5.2 and 5.4.

The GMANOVA model and the problem, along with some examples, are presented in Section 5.2. The examples bring out the subtle differences between GMANOVA and MANOVA problems. A historical development of the solution to the GMANOVA problem is also discussed in this section. Finally, canonical forms of GMANOVA, MANOVA, and ANOVA, mentioned in Section 3.2 of Chapter 3 and used throughout that chapter, are derived here.

Section 5.3 deals with the MANOVA problem in a canonical setup (see (3.10) in Section 3.2 of Chapter 3). The optimality robustness of the UMPI property of Pillai's test in a suitable class of distributions (similar to Example 3.5 in Section 3.5 of Chapter 3) under the restriction on the dimension of the parameters or the observations that $\min(n_1, p) = 1$ is proved by using Theorem 3.11 of Chapter 3 (Theorem 5.1). This covers as special cases the optimality robustness of Hotelling's T^2-test, the ANOVA F-test, and the likelihood ratio test (LRT) in a typical case of the MANOVA. The optimality robustness of the LBI test of Pillai in general when $\min(n_1, p) > 1$ is deferred to Section 5.4 (see Corollary 5.3).

The GMANOVA problem is discussed in detail in Section 5.4. The optimality robustness of the LBI test in general within a suitable class of distributions is established by following the general procedure developed in Section 3.5 of Chapter 3 (Theorem 5.3). As a special case, we obtain the optimality robustness of the LBI test of Pillai for the general MANOVA problem when $\min(n_1, p) > 1$. It may be mentioned that Schwartz (1967a) first proved that Pillai's test for MANOVA is LBI under the assumption of normality of the observations. The LRT and some related tests are briefly discussed in Remark 5.3 and Remark 5.4.

5.2. GMANOVA Model and Problem

As mentioned in Section 3.2 of Chapter 3, the GMANOVA model refers to an observed data matrix $Y: n \times p$ obeying the probability model

$$Y = R_1 B R_2 + E, \tag{5.1}$$

where $R_1: n \times (n_1 + n_2)$ of rank $(n_1 + n_2)$, $R_2: (p_1 + p_2) \times p$ of rank $(p_1 + p_2)$ are known matrices, $B: (n_1 + n_2) \times (p_1 + p_2)$ is a matrix of unknown parameters, and the error matrix $E: n \times p$ is assumed to follow

$$E \sim \mathscr{F}_{\mathrm{E}}(0, I_n \otimes \Omega) \tag{5.2}$$

for some unknown $\Omega \in \mathscr{S}(p)$. Here $n = n_1 + n_2 + n_3$, $p = p_1 + p_2 + p_3$, and $n_3 \geqq p$ is assumed. It is noted that this model is a version of (3.5) in Section 3.2 of Chapter 3 in that $\mu \in R^{n \times p}$ is taken to be of the special form $R_1 B R_2$ given above, Φ is taken as I_n, and normality there is replaced by the class of elliptically symmetric densities depicted in (5.2) above. Special forms of the model (5.1) have been considered in various problems in association with growth curves. The above general formulation is due to Potthoff and Roy (1964), though E there is assumed to be normal. The problem of estimating the coefficient matrix B in the model (5.1), which is not the subject matter of this book, is treated in Rao (1965, 1967), Geisser (1970). Throughout this chapter, we are concerned with the problem of testing a general linear hypothesis of the form

$$\text{(5.3)} \qquad \text{H}: R_3 B R_4 = R_0 \qquad \text{versus} \qquad \text{K}: R_3 B R_4 \neq R_0$$

under the models (5.1)–(5.2), where $R_3: n_1 \times (n_1 + n_2)$ of rank n_1, R_4: $(p_1 + p_2) \times p_2$ of rank p_2, and $R_0: n_1 \times p_2$ are known matrices. This testing problem is often called the GMANOVA problem, and the hypothesis (5.3) is usually referred to as a GMANOVA hypothesis.

Under the assumption of normality of E, the GMANOVA problem is treated in Khatri (1966), Gleser and Olkin (1970), Kariya (1978); some of the problems associated with the GMANOVA model are considered by Hooper (1982, 1983), and Marden (1983). The analysis of the GMANOVA problem via invariance under the distributional assumption (5.2) of E is the main theme of this chapter.

An important special case of the GMANOVA problem corresponds to the case when $R_2 = R_4 = I$ and $p_1 = p_3 = 0$; this is the well-known MANOVA problem. It is stated in terms of the model

$$\text{(5.4)} \qquad Y = R_1 B + E$$

and the hypothesis

$$\text{(5.5)} \qquad \text{H}: R_3 B = R_0 \qquad \text{versus} \qquad \text{K}: R_3 B \neq R_0$$

where the error matrix E follows the law in (5.2). As mentioned before, this special setup covers plenty of problems, e.g., problems of testing on means, regression coefficients, effects of treatments, etc.

Some examples of the GMANOVA problem that are not special cases of the MANOVA problem are given below. The reader is referred to the book by Kariya (1985) in this connection.

Example 5.1. (Potthoff and Roy, 1964). Let $y(t) = (y_1(t), \ldots, y_p(t))'$ be a $p \times 1$ vector consisting of p economic variables or growth characteristics. Assume that $y(t)$ is observed at time $t = 1, 2, \ldots, n$, resulting in the data matrix $Y = (y(1), \ldots, y(n))'$: $n \times p$. By denoting the mean of $y(t)$ by $\mu(t)$, it is further assumed that the ith element $\mu_i(t)$ of $\mu(t)$ is a polynomial in t of degree $(k - 1)$

given by

$$\mu_i(t) = \beta_{io} + \beta_{i1}t + \cdots + \beta_{ik-1}t^{k-1}, \qquad i = 1,\ldots,p. \tag{5.6}$$

This is a model in which p economic time series or growth characteristics $(y_1(t),\ldots,y_p(t))$ fluctuate at each time t, and $y_i(t)$ fluctuates around the polynomial $\mu_i(t)$. The model can be written as a MANOVA model (5.4), where

$$Y = \begin{bmatrix} y(1)' \\ \vdots \\ y(n)' \end{bmatrix}, \qquad R_1 = \begin{bmatrix} 1 & 1 & 1 & \cdots & 1 \\ 1 & 2 & 2^2 & \cdots & 2^{k-1} \\ \vdots & \vdots & \vdots & & \vdots \\ 1 & n & n^2 & \cdots & n^{k-1} \end{bmatrix} : n \times k \tag{5.7}$$

and

$$B = \begin{bmatrix} \beta_{10} & \beta_{20} & \cdots & \beta_{po} \\ \beta_{11} & \beta_{21} & \cdots & \beta_{p1} \\ \hdashline \beta_{1k-1} & \beta_{2k-1} & \cdots & \beta_{pk-1} \end{bmatrix} : k \times p. \tag{5.8}$$

In this model the null hypothesis that the p underlying variables follow the same growth pattern completely is expressed as

$$\text{H}: \beta_{1j} = \beta_{2j} = \cdots = \beta_{pj}, \qquad j = 0, 1, \ldots, k-1, \tag{5.9}$$

which is of the form (5.3) with $R_3 = I, R_0 = 0$ and

$$R_4 = \begin{bmatrix} 1 & 0 & 0 & \cdots & 0 \\ -1 & 1 & 0 & \cdots & 0 \\ 0 & -1 & 1 & \cdots & 0 \\ \vdots & \vdots & \vdots & & \vdots \\ 0 & 0 & 0 & \cdots & 1 \\ 0 & 0 & 0 & \cdots & -1 \end{bmatrix} : p \times (p-1). \tag{5.10}$$

Moreover, the hypothesis that the p underlying variables follow the same growth pattern except for original levels or intercepts β_{io}(at $t = 0$) is expressed again in the form (5.3) with $R_0 = 0$, $R_3 = [0, I_{k-1}]$: $(k-1) \times k$, and R_4, as given in (5.10). This represents the null hypothesis

$$\text{H}: \beta_{1j} = \cdots = \beta_{pj} \qquad (j = 1,\ldots,k-1), \tag{5.11}$$

but the β_{io}'s may be different. Thus, though the model (5.6) is a MANOVA model (5.4), the hypothesis (5.9) or (5.11) is a GMANOVA hypothesis (5.3) different from a MANOVA hypothesis (5.5) because the coefficient matrix B is multiplied by the matrix R_4 from the right. Therefore, the problem of testing (5.9) or (5.11) is not in the framework of the MANOVA problem but rather a special case of the GMANOVA problem.

***Example* 5.2.** Suppose there are two groups of states where the first group consists of k_1, eastern states, and the second, of k_2, western states. It is judged that the different states in each group have experienced the same growth process in terms of economy (or some other characteristic). Here the problem is to test the hypothesis that the two groups have the same growth pattern based on p economic (or similar) variables of the same kind for each group.

Data consist of $y_{1j}(t)=(y_{1j1}(t), y_{1j2}(t), \ldots, y_{1jp}(t))'$, $j=1,\ldots,k_1$, $t=1,\ldots,n$, and $y_{2j}(t)=(y_{2j1}(t), y_{2j2}(t), \ldots, y_{2jp}(t))'$, $i=1,\ldots,k_2$, $t=1,\ldots,n$, representing values of the p variables over n points of time for each state in both the groups. Let

$$(5.12)\qquad Y_1 = \begin{bmatrix} y'_{11}(1) & y'_{12}(1) & \cdots & y'_{1k_1}(1) \\ y'_{11}(2) & y'_{12}(2) & \cdots & y'_{1k_1}(2) \\ \cdots & \cdots & \cdots & \cdots \\ y'_{11}(n) & y'_{12}(n) & \cdots & y'_{1k_1}(n) \end{bmatrix} : n \times k_1 p,$$

$$Y_2 = \begin{bmatrix} y'_{21}(1) & y'_{22}(1) & \cdots & y'_{2k_2}(1) \\ y'_{21}(2) & y'_{22}(2) & \cdots & y'_{2k_2}(2) \\ \cdots & \cdots & \cdots & \cdots \\ y'_{21}(n) & y'_{22}(n) & \cdots & y'_{2k_2}(n) \end{bmatrix} : n \times k_2 p.$$

Since we have assumed that all the states in each group follow the same growth process, we have

$$(5.13)\qquad \begin{aligned} E(y_{11}(t)) &= \cdots = E(y_{1k_1}(t)) = \mu_1(t) \\ E(y_{21}(t)) &= \cdots = E(y_{2k_2}(t)) = \mu_2(t) \end{aligned} \qquad t = 1,\ldots,n.$$

Furthermore, we assume that $\mu_1(t)$ and $\mu_2(t)$ are approximated by polynomials of degree $(k-1)$. Then, as in Example 5.1, define B_1 and B_2 for each group as in (5.8), and use the same R_1 as in (5.7) to obtain

$$(5.14)\qquad \begin{aligned} E(Y_1) &= (R_1B_1, R_1B_1, \ldots, R_1B_1) = R_1B_1\overbrace{[I_p, I_p, \ldots, I_p]}^{k_1} \\ E(Y_2) &= (R_1B_2, R_1B_2, \ldots, R_1B_2) = R_1B_2\underbrace{[I_p, I_p, \ldots, I_p]}_{k_2} \end{aligned}$$

Hence, by writing

$$(5.15) \qquad Y = [Y_1 : Y_2]: n \times (k_1 + k_2)p, \qquad B = [B_1 : B_2]: k \times 2p$$

and

$$(5.16) \qquad R_2 = \begin{bmatrix} \overbrace{I_p \quad I_p \ \dots \ I_p}^{k_1} & \overbrace{0 \quad 0 \quad \dots \quad 0}^{k_2} \\ 0 \quad 0 \ \dots \ 0 & I_p \quad I_p \quad \dots \quad I_p \end{bmatrix} : 2p \times (k_1 + k_2)p,$$

our model is finally expressed as

$$(5.17) \qquad Y = R_1 B R_2 + E,$$

E following (5.2), where all the p variables in the $(k_1 + k_2)$ states are assumed to have the same scale matrix Ω. The hypothesis to be tested is $B_1 = B_2$, and so it is expressed as

$$(5.18) \qquad BR_4 = 0 \qquad \text{with} \quad R_4 = \begin{bmatrix} I_p \\ -I_p \end{bmatrix} : 2p \times p.$$

The other hypothesis, namely, that the growth pattern between the two groups is the same except for initial levels or intercepts, is of the form $R_3 B R_4 = 0$ with $R_3 = [0, I_{k-1}]: k - 1 \times k$.

These are examples of GMANOVA problems that pool cross-section data and time series data.

Example 5.3. Let $x_i = (x'_{1i}, x'_{2i})': 2p \times 1$, $x_{ki}: p \times 1$, $k = 1, 2, i = 1, \dots, n$, $\mu = (\mu'_1, \mu'_2)': 2p \times 1$, $\mu_k: p \times 1$, $k = 1, 2$. Then letting $Y' = [x_1, \dots, x_n]$ and $\underset{\sim}{1} = (1, \dots, 1)': n \times 1$, consider the model

$$(5.19) \qquad Y = R_1 B + E \qquad \text{with} \quad R_1 = \underset{\sim}{1}, \quad B = \mu',$$

E following (5.2). This is a MANOVA model, but the hypothesis of interest $\mu_1 = \mu_2$ is expressed as

$$(5.20) \qquad BR_4 = 0 \qquad \text{with} \quad R_4 = \begin{bmatrix} I_p \\ -I_p \end{bmatrix}.$$

Hence, testing this hypothesis is not a special case of the MANOVA problem but a special case of the GMANOVA problem.

Example 5.4. A generalization of the preceding example consists of observing $x_i = (x'_{1i}, x'_{2i})': (p_1 + p_2) \times 1$, $x_{ki}: p_k \times 1$, $k = 1, 2, i = 1, \dots, n_1$ and $z_j = (z'_{1j}, z'_{2j})': (p_1 + p_2) \times 1$, $z_{kj}: p_k \times 1$, $k = 1, 2, j = 1, \dots, n_2$. Let $\mu =$

$(\mu_1', \mu_2')'$: $(p_1 + p_2) \times 1$, $\eta = (\eta_1', \eta_2')'$: $(p_1 + p_2) \times 1$, μ_k: $p_k \times 1$, η_k: $p_k \times 1$, $k = 1, 2$, and

$$Y = [x_1, \ldots, x_{n_1}, z_1, \ldots, z_{n_2}]' : (n_1 + n_2) \times (p_1 + p_2)$$

$$B = \begin{bmatrix} \mu_1' & \mu_2' \\ \eta_1' & \eta_2' \end{bmatrix} : 2 \times (p_1 + p_2), \qquad R_1 = \begin{bmatrix} \underset{\sim}{1}_1 & 0 \\ 0 & \underset{\sim}{1}_2 \end{bmatrix} : (n_1 + n_2) \times 2 \tag{5.21}$$

$$(\underset{\sim}{1}_i = (1, \ldots, 1)' : n_i \times 1, \qquad i = 1, 2).$$

Then the model is expressed as

$$Y = R_1 B + E, \tag{5.22}$$

E following (5.2), and the hypothesis of interest $\mu_2 = \eta_2$ as

(5.23)

$$R_3 B R_4 = 0, \qquad R_3 = (1, -1) : 1 \times 2, \qquad R_4 = \begin{bmatrix} 0 \\ I_{p_2} \end{bmatrix} : (p_1 + p_2) \times p_2.$$

This is a special case of the GMANOVA problem.

Some useful extensions of the GMANOVA problem are discussed in Gleser and Olkin (1970) and Banken (1984). The reader is referred to the monograph by Kariya (1985) for details.

In contrast to the usual MANOVA problem, very little work was done on the GMANOVA problem until the seventies. References to work on this model prior to that of Potthoff and Roy (1964) and Rao (1965) can be found in Rao (1967) and Gleser and Olkin (1970). Khatri (1966) derived the likelihood ratio test (LRT) for the above problem by suitably transforming the model and then using a conditional argument. He also proposed some other related tests based on similarities between this problem and the MANOVA problem (see Remark 5.4 below). By using Khatri's approach, Krishnaiah (1966) studied some related problems in this model. Gleser and Olkin (1970), on the other hand, reduced the problem to a very convenient canonical form and used the invariance principle to derive the LRT. Kiefer and Schwartz (1965) briefly treated the problem in a special case and proposed a noninvariant Bayes test. In the same case, Stein (1966) proposed a conditional test based on the principle of conditionality. Fujikoshi (1973) proved the unconditional monotonicity of the power functions of the LRT and the tests proposed by Khatri (1966) and also derived the nonnull asymptotic distributions of these tests. Geisser (1970, 1980), Lee and Geisser (1972, 1975), and Fearn (1975) made Bayesian analyses of the GMANOVA problem. All of the above works are done under the assumption of the normality of E.

The first systematic study of the GMANOVA problem, using the invariance approach, was undertaken by Kariya (1978), who derived the locally best invariant (LBI) test and showed that it is locally minimax. In the same year, Kariya and Kanazawa (1978) derived the null distribution of the LBI test for a special case. The admissibility of tests in the GMANOVA problem is questioned in Marden and Perlman (1980) and Marden (1983), and complete classes are described. In particular, the LRT is shown to be inadmissible. The problem of simultaneous interval estimation in the GMANOVA model is discussed in Hooper (1982, 1983). In applications, Ware and Bowden (1977) have applied the model to a circadian rhythm analysis, and Zerbe and Jones (1980) to a time series analysis.

Again, all of the above work was done under the assumption that E is normal. Recently, however, Kariya and Sinha (1985) established the optimality robustness of the LBI test of Kariya (1978) for the GMANOVA problem for a broad class of q in $\mathscr{F}_E$ containing the normal, using the powerful tool of Wijsman's representation theorem (Chapter 2). In particular, this establishes the optimality robustness of the LBI test of Schwartz (1967a) for the MANOVA problem when $\min(n_1, p) > 1$. Earlier, Kariya (1981a) established the optimality robustness of the UMPI test for the MANOVA problem under the condition that $\min(n_1, p) = 1$ within a certain class of distributions of E somewhat broader than $\mathscr{F}_{\mathrm{E}}$. This result is presented in the next section by using Theorem 3.11 of Chapter 3.

The analysis of the GMANOVA problem as formulated in (5.1)–(5.3) and that of the MANOVA problem as described in (5.4)–(5.5) is based on the reduction to a canonical form that is developed below.

Canonical form. We use the following elementary result for a canonical reduction.

Lemma 5.1. *For any $m \times n$ matrix A with rank $(A) = m$, there exist matrices $F \in \mathrm{GL}(m)$ and $P \in \mathcal{O}(n)$ such that $A = F(I_m, 0)P$.*

By this lemma, we can write R_1 and R_2 as

(5.24)
$$R_1 = P_1\begin{pmatrix} I_{n_1+n_2} \\ 0 \end{pmatrix}F_1, \qquad P_1 \in \mathcal{O}(n), \qquad F_1 \in \mathrm{GL}(n_1 + n_2),$$
$$R_2 = F_2(I_{p_1+p_2}, 0)P_2, \qquad P_2 \in \mathcal{O}(p), \qquad F_2 \in \mathrm{GL}(p_1 + p_2).$$

Let

(5.25)
$$Y^* = P_1' Y P_2', \qquad B^* = F_1 B F_2, \qquad E^* = P_1' E P_2', \qquad \Omega^* = P_2' \Omega P_2.$$

Then the model (5.1) reduces to

$$Y^* = \begin{pmatrix} B^* & 0 \\ 0 & 0 \end{pmatrix} + E^* \tag{5.26}$$

and the distribution (5.2) to

$$E^* \sim \mathscr{F}_E(0, I_n \otimes \Omega^*). \tag{5.27}$$

On the other hand, for the null and alternative hypotheses in (5.3), we can write $R_3 B R_4$ as

$$R_3 B R_4 = R_3 F_1^{-1} B^* F_2^{-1} R_4 \tag{5.28}$$

and apply Lemma 5.1 to $R_3 F_1^{-1}$ and $F_2^{-1} R_4$ to get

$$\begin{aligned} R_3 F_1^{-1} &= F_3(I_{n_1}, 0) P_3, \qquad F_3 \in \mathrm{GL}(n_1), \qquad P_3 \in \mathcal{O}(n_1 + n_2), \\ F_2^{-1} R_4 &= P_4 \begin{pmatrix} 0 \\ I_{p_2} \end{pmatrix} F_4, \qquad F_4 \in \mathrm{GL}(p_2), \qquad P_4 \in \mathcal{O}(p_1 + p_2). \end{aligned} \tag{5.29}$$

Hence, from (5.28) and (5.29), the hypotheses in (5.3) can be written as

$$\begin{aligned} &\mathrm{H}: (I_{n_1}, 0) P_3 B^* P_4 \begin{pmatrix} 0 \\ I_{p_2} \end{pmatrix} = F_3^{-1} R_0 F_4^{-1} \qquad \text{versus} \\ &\mathrm{K}: (I_{n_1}, 0) P_3 B^* P_4 \begin{pmatrix} 0 \\ I_{p_2} \end{pmatrix} \neq F_3^{-1} R_0 F_4^{-1}. \end{aligned} \tag{5.30}$$

Set

$$\Theta = P_3 B^* P_4 : (n_1 + n_2) \times (p_1 + p_2), \; \bar{\Theta} = (\Theta, 0) : (n_1 + n_2) \times p,$$

$$\tilde{\Theta} \equiv \begin{pmatrix} \bar{\Theta} \\ 0 \end{pmatrix} \equiv \begin{pmatrix} \Theta & 0 \\ 0 & 0 \end{pmatrix} = \begin{array}{c} \begin{matrix} p_1 & \;\; p_2 & \;\; p_3 \end{matrix} \\ \begin{pmatrix} \Theta_{11} & \Theta_{12} & \Theta_{13} \\ \Theta_{21} & \Theta_{22} & \Theta_{23} \\ 0 & 0 & 0 \end{pmatrix} \end{array} \begin{matrix} n_1 \\ n_2 \\ n_3 \end{matrix} \quad \text{with} \quad \begin{matrix} \Theta_{13} \equiv 0, \\ \Theta_{23} \equiv 0. \end{matrix} \tag{5.31}$$

Without loss of generality, we assume $R_0 = 0$. Now define X and Σ, respectively, by

$$\begin{aligned} X &= \begin{pmatrix} P_3 & 0 \\ 0 & I \end{pmatrix} Y^* \begin{pmatrix} P_4 & 0 \\ 0 & I \end{pmatrix} : n \times p, \\ \Sigma &= \begin{pmatrix} P_4' & 0 \\ 0 & I \end{pmatrix} \Omega^* \begin{pmatrix} P_4 & 0 \\ 0 & I \end{pmatrix} : p \times p. \end{aligned} \tag{5.32}$$

Then the canonical form of the models (5.1)–(5.2) is expressed as

$$
(5.33) \qquad X \equiv \begin{pmatrix} \overset{p_1}{X_{11}} & \overset{p_2}{X_{12}} & \overset{p_3}{X_{13}} \\ X_{21} & X_{22} & X_{23} \\ X_{31} & X_{32} & X_{33} \end{pmatrix} \begin{matrix} n_1 \\ n_2 \\ n_3 \end{matrix} = \begin{pmatrix} \Theta_{11} & \Theta_{12} & 0 \\ \Theta_{21} & \Theta_{22} & 0 \\ 0 & 0 & 0 \end{pmatrix} + \tilde{E},
$$

where

$$
(5.34) \qquad \tilde{E} = \begin{pmatrix} P_3 & 0 \\ 0 & I \end{pmatrix} E * \begin{pmatrix} P_4 & 0 \\ 0 & I \end{pmatrix} : n \times p \sim \mathscr{F}_E(0, I_n \otimes \Sigma),
$$

and the problem is to test

$$(5.35) \qquad \text{H: } \Theta_{12} = 0 \quad \text{versus} \quad \text{K: } \Theta_{12} \neq 0.$$

This is precisely what is given in (3.8)–(3.9) in Section 3.2 of Chapter 3 under the assumption that $\tilde{E}$ is normal.

Here we may further reduce the models (5.33)–(5.34) by sufficiency to the consideration of

$$(5.36) \qquad \bar{X} = \begin{pmatrix} X_{11} & X_{12} & X_{13} \\ X_{21} & X_{22} & X_{23} \end{pmatrix} \quad \text{and} \quad V = (V_{ij}) = (X'_{3i} X_{3j}): p \times p.$$

But the sufficient statistic $(\bar{X}, V)$ is not complete because we have the extra information $\Theta_{13} = 0$, $\Theta_{23} = 0$. When q in $\mathscr{F}_E$ is exponential, $\bar{X} \sim N(\bar{\Theta}, I_{n_1 + n_2} \otimes \Sigma)$, $V \sim W_p(\Sigma, n_3)$, and $\bar{X}$, V are independent. Because of the information $\Theta_{13} = 0$, $\Theta_{23} = 0$, the joint distribution of $\bar{X}$ and V is curved exponential, which in terms of the natural parameters $\Delta \equiv \Sigma^{-1}$ and $\eta \equiv \bar{\Theta}\Sigma^{-1}$ can be written as

$$(5.37) \qquad f(\bar{x}, v \mid \Delta, \eta) = \beta(\Delta, \eta) h(v) \exp[-\tfrac{1}{2} \operatorname{tr} \Delta(\bar{x}'\bar{x} + v) + \operatorname{tr} \bar{x}'\eta].$$

Unfortunately, however, neither the information $\Theta_{13} = 0$, $\Theta_{23} = 0$ on $\bar{\Theta}$ nor the hypothesis H: $\Theta_{12} = 0$ is linear in the natural parameters Δ and η. This nonlinearity of the model as well as of the hypothesis makes the analysis of the GMANOVA problem difficult. Although the above interpretation of a curved exponential family is not there when q is not exponential, the difficulty in the analysis prevails.

A special case of the GMANOVA problem, as mentioned before, is the MANOVA problem that corresponds to $p_1 = p_3 = 0$ in (5.33). Thus, in canonical form, the MANOVA model (5.4) can be written as (letting $p_2 = p$)

$$X: n \times p \equiv \overset{p}{\begin{pmatrix} X_1 \\ X_2 \\ X_3 \end{pmatrix}} \begin{matrix} n_1 \\ n_2 \\ n_3 \end{matrix} = \begin{pmatrix} \mu_1 \\ \mu_2 \\ 0 \end{pmatrix} + \tilde{E}, \qquad \tilde{E}: n \times p \sim \mathscr{F}_{\mathrm{E}}(0, I_n \otimes \Sigma) \tag{5.38}$$

and the MANOVA hypothesis (5.5) as

$$\mathrm{H}: \mu_1 = 0 \qquad \text{versus} \qquad \mathrm{K}: \mu_1 \neq 0, \qquad \mu_2 \in R^{n_2 \times p}, \tag{5.39}$$

with $\Sigma \in \mathscr{S}(p)$ being unknown. The reader may note that this is the canonical form that appears in (3.10) in Section 3.2 of Chapter 3. The apparent simplicity of the MANOVA problem compared to the GMANOVA problem is evident from the above discussion.

Before we analyse the MANOVA problem in its canonical form in the next section, a correspondence between the original observation matrix Y and its canonical analogue X is in order. Let $\tilde{R}_1$ and $\tilde{R}_2$, respectively, be $n \times n_3$ and $p_3 \times p$ matrices such that

$$\begin{aligned} &\operatorname{rank}[R_1, \tilde{R}_1] = n, \qquad \tilde{R}_1' R_1 = 0, \\ &\operatorname{rank}\begin{bmatrix} R_2 \\ \tilde{R}_2 \end{bmatrix} = p, \qquad R_2 \tilde{R}_2' = 0. \end{aligned} \tag{5.40}$$

Similarly, let $\tilde{R}_3: n_2 \times (n_1 + n_2)$ and $\tilde{R}_4: (p_1 + p_2) \times p_1$ satisfy

$$\begin{aligned} &\operatorname{rank}\begin{bmatrix} R_3 \\ \tilde{R}_3 \end{bmatrix} = n_1 + n_2, \qquad \mathrm{R}_3 \tilde{R}_3' = 0, \\ &\operatorname{rank}[\tilde{R}_4, R_4] = p_1 + p_2, \qquad \tilde{R}_4' R_4 = 0. \end{aligned} \tag{5.41}$$

Then for the P's and F's in (5.24) and (5.29) we can take

(5.42)

$$F_1 = (R_1' R_1)^{1/2}, \qquad F_2 = (R_2 R_2')^{1/2}, \qquad P_1 = [R_1 (R_1' R_1)^{-1/2}, \tilde{R}_1 (\tilde{R}_1' \tilde{R}_1)^{-1/2}]$$

$$P_2 = \begin{pmatrix} (R_2 R_2')^{-1/2} R_2 \\ (\tilde{R}_2 \tilde{R}_2')^{-1/2} \tilde{R}_2 \end{pmatrix},$$

$$\begin{aligned} &F_3 = (R_3 F_1^{-2} R_3')^{1/2}, \qquad F_4 = (R_4' F_2^{-2} R_4)^{1/2}, \\ &P_3 = \begin{pmatrix} (R_3 F_1^{-2} R_3')^{-1/2} R_3 F_1^{-1} \\ (\tilde{R}_3 F_1^{-2} \tilde{R}_3')^{-1/2} \tilde{R}_3 F_1 \end{pmatrix}, \\ &P_4 = [F_2 \tilde{R}_4 (\tilde{R}_4' F_2^{-2} \tilde{R}_4)^{-1/2}, F_2^{-1} R_4 (R_4' F_2^{-2} R_4)^{-1/2}]. \end{aligned} \tag{5.43}$$

By using these expressions in (5.25) and (5.32), the canonical variables X_{ij}'s are written in terms of the original variables Y_{ij}'s. In particular, we get the following:

$$
\begin{aligned}
X_{12} &= [R_3(R_1'R_1)^{-1}R_3']^{-1/2}R_3\hat{B}_0R_4[R_4'(R_2R_2')^{-1}R_4]^{-1/2},\\
\hat{B}_0 &= (R_1'R_1)^{-1}R_1'YR_2'(R_2R_2')^{-1},\\
X_{13} &= [R_3(R_1'R_1)^{-1}R_3']^{-1/2}R_3(R_1'R_1)^{-1}R_1'Y\tilde{R}_2'(\tilde{R}_2\tilde{R}_2')^{-1/2},\\
X_{32} &= (\tilde{R}_1'\tilde{R}_1)^{-1/2}R_1'YR_2'(R_2R_2')^{-1}R_4[R_4'(R_2R_2')^{-1}R_4]^{-1/2},\\
X_{33} &= (\tilde{R}_1'\tilde{R}_1)^{-1/2}\tilde{R}_1'Y\tilde{R}_2'(\tilde{R}_2\tilde{R}_2')^{-1/2}.
\end{aligned}
\tag{5.44}
$$

It may be noted that

$$
\begin{aligned}
\tilde{R}_1(\tilde{R}_1'\tilde{R}_1)^{-1}\tilde{R}_1' &= I_n - R_1(R_1'R_1)^{-1}R_1',\\
\tilde{R}_2'(\tilde{R}_2\tilde{R}_2')^{-1}\tilde{R}_2 &= I_p - R_2'(R_2R_2')^{-1}R_2.
\end{aligned}
\tag{5.45}
$$

5.3. MANOVA Problem

The MANOVA problem in its canonical form (5.38)–(5.39) has been briefly discussed in Sections 3.2, 3.3, and 3.5 of Chapter 3. As mentioned in the Introduction in Section 5.1, we here establish the optimality robustness of the UMPI property of Pillai's test based on tr $X_1'X_1(X_1'X_1 + X_3'X_3)^{-1}$ with the help of Theorem 3.11 of Chapter 3 under the restriction min $(n_1, p) = 1$.

In what follows, we work with a somewhat wider framework than (5.38) in that the distributional assumption $\tilde{E} \sim \mathscr{F}_{\mathrm{E}}(0, I_n \otimes \Sigma)$ is extended to include more probability distributions. Let Q be the class of functions from the set of $p \times p$ matrices into $[0, \infty)$ such that for $q \in Q$,

$$
q \text{ is convex on } \bar{\mathscr{S}}(p), \qquad \int_{R^{n\times p}} q(x'x)\,dx = 1,
\tag{5.46}
$$

and

$$
q(BV) = q(VB) \qquad \text{for all} \qquad V \in \bar{\mathscr{S}}(p), \qquad B \in \mathrm{GL}(p).
$$

In the above, $\bar{\mathscr{S}}(p)$ denotes the set of $p \times p$ nonnegative-definite matrices. Our extended model then consists of the same framework as in (5.38), except that the error term $\tilde{E}$ is assumed to have a density of the form

$$
\tilde{E} \sim f(\tilde{E}\,|\,\Sigma) = |\Sigma|^{-n/2}q(\Sigma^{-1}\tilde{E}'\tilde{E}), \qquad q \in Q.
\tag{5.47}
$$

Note that in (5.47) we have spelled out clearly what conditions we require of q, namely $q \in Q$, in order that a UMPI test of the MANOVA hypothesis (5.39) exists under the condition $\min(n_1, p) = 1$. Of course, when $p = 1$, Q coincides with $\mathscr{F}_E$ when convexity is imposed on the members of the latter. The condition (5.46) is trivially satisfied when q is either normal or a normal scale mixture. (Exercise 5.1)

In order to derive an optimum invariant test, note that the testing problem (5.39) under the above MANOVA model is invariant under the group $\mathscr{G} = \mathscr{O}(n_1) \times \mathscr{O}(n_2) \times \mathscr{O}(n_3) \times \mathrm{GL}(p) \times R^{n_2 \times p}$ acting on x by

$$(5.48) \qquad x \to g(x) = \begin{pmatrix} Q_1 & 0 & 0 \\ 0 & Q_2 & 0 \\ 0 & 0 & Q_3 \end{pmatrix} xC' + \begin{pmatrix} 0 \\ F \\ 0 \end{pmatrix}$$

with $g = (Q_1, Q_2, Q_3, C, F) \in \mathscr{G}$. The group $\mathscr{G}$ and its action on x, namely (5.48), already appear in (3.11) and (3.12), respectively, of Chapter 3 when the error term is normal. They remain unchanged when normality is replaced by (5.47). Denote a maximal invariant under $\mathscr{G}$ by $\pi(x)$. The probability ratio R of its nonnull to null distribution is derived below. See (3.47) and (3.48) in Section 3.4 of Chapter 3 providing an expression of R. Note that here Θ plays the role of θ so that Θ_0 and Θ_1 represent, respectively, points in the null and the alternative parameter spaces.

Lemma 5.2. *When* $\min(n_1, p) = 1$, *the ratio* $R = \dfrac{dP^{\pi}_{\Theta_1}}{dP^{\pi}_{\Theta_0}}(\pi(x))$ *is given by*

$$R = R(\pi(x))$$

$$(5.49) \qquad = \frac{\int_{\mathrm{GL}(p) \times \mathscr{O}(n_1)} \tilde{q}(AA' + \delta \underset{\sim}{e}\underset{\sim}{e}' - v\delta^{1/2} r_{11}(\underset{\sim}{a}_1 \underset{\sim}{e}' + \underset{\sim}{e}\underset{\sim}{a}'_1)) |AA'|^{k/2}\, dA\, dv(Q_1)}{\int_{\mathrm{GL}(p)} \tilde{q}(AA') |AA'|^{k/2}\, dA}$$

where $\tilde{q}(v) = \int_{R^{n_2 \times p}} q(v + x'x)\, dx$, $v = T/(1+T)^{1/2}$ *with* $T = \mathrm{tr}\, X_1 (X'_3 X_3)^{-1} X'_1$, $k = n_1 + n_3 - p$, $\underset{\sim}{e} = (1, 0, \ldots, 0)' \in R^p$, $\delta = \mathrm{tr}\, \mu_1 \Sigma^{-1} \mu'_1$, $\underset{\sim}{a}_i$ *is the ith column of* A, r_{ij} *is the* (i, j)*th element of* Q_1, *and* $dv(Q_1)$ *is the invariant probability measure over* $\mathscr{O}(n_1)$.

Proof. Choose $v(dg) = dv(Q_1) \cdot dv(Q_2) \cdot dv(Q_3) \cdot d\mu(C) \cdot dF$ as a left-invariant measure on $\mathscr{G}$ where $dv(Q_i)$ is the invariant probability measure on $\mathscr{O}(n_i)$, $d\mu(C) = dC/|CC'|^{p/2}$ with dC Lebesgue on R^{p^2}, and dF is Lebesgue on $R^{n_2 \times p}$. Using (5.48) and the fact that $\chi(g)$, the inverse of the Jacobian of transformation $x \to g(x)$, is simply $|CC'|^{(n_1+n_3)/2}$, the numerator $H(x \mid \Theta_1)$ of (3.48) of Chapter 3 is expressed as

(5.50)

$$
\begin{aligned}
H(x \mid \Theta_1) &= \int_{\mathscr{G}} f(g(x) \mid \Theta)\chi(g)\nu(dg) \\
&= \int_{\mathscr{G}} q[\Sigma^{-1}(Q_1 X_1 C' - \mu_1)'(Q_1 X_1 C' - \mu_1) \\
&\quad + \Sigma^{-1}(Q_2 X_2 C' + F - \mu_2)'(Q_2 X_2 C' + F - \mu_2) \\
&\quad + \Sigma^{-1} C X_3' X_3 C']|CC'|^{(n_1+n_3)/2} d\nu(Q_1)\, d\nu(Q_2)\, d\nu(Q_3)\, d\mu(C)\, dF.
\end{aligned}
$$

Using the property $q(BV) = q(VB)$ of q in (5.46), and integrating out F over $R^{n_2 \times p}$ with respect to dF yields

(5.51)

$$
\begin{aligned}
H(x \mid \Theta_1) = \int_{\mathcal{O}(n_1) \times \mathrm{GL}(p)} & \tilde{q}[\Sigma^{-1/2}(Q_1 X_1 C' - \mu_1)'(Q_1 X_1 C' - \mu_1)\Sigma^{-1/2} \\
& + \Sigma^{-1/2} C X_3' X_3 C' \Sigma^{-1/2}]|CC'|^{(n_1+n_3)/2}\, d\mu(C)\, d\nu(Q_1).
\end{aligned}
$$

Since μ is both left and right invariant, replacing C by $C^* = \Sigma^{-1/2} C (X_3' X_3)^{1/2}$ leaves the integral in (5.51) the same. Write $c_1 = |X_3' X_3|^{-(n_1+n_3)/2}$, $u = X_1 (X_3' X_3)^{-1/2}$: $n_1 \times p$, and $\eta = \Sigma^{-1/2}\mu_1'$: $p \times n_1$. Note that $n_3 \geqq p$ implies $(X_3' X_3)^{-1/2} \in \mathscr{S}(p)$. Making the above transformation from C to C^* reduces $H(x \mid \Theta_1)$ to

(5.52)

$$
\begin{aligned}
H(x \mid \Theta_1) &= |\Sigma|^{(n_1+n_3)/2} \cdot c_1 \\
&\quad \cdot \int_{\mathcal{O}(n_1) \times \mathrm{GL}(p)} \tilde{q}(C^* u' u C^{*\prime} - C^* u' Q_1' \eta' - \eta Q_1 u C^{*\prime} + \eta\eta' + C^* C^{*\prime}) \\
&\qquad \times |C^* C^{*\prime}|^{k/2}\, dC^*\, d\nu(Q_1),
\end{aligned}
$$

where $k = n_1 + n_3 - p$. Of course, in our problem, $H(x \mid \Theta_0)$, which is the denominator of R (see (3.47) in Section 3.4 of Chapter 3), is obtained from the right-hand side of (5.52) by putting $\mu_1 = 0$, i.e., $\eta = 0$.

The derivation so far is quite general. Now assume that $n_1 = 1$ and let U_1, U_2 be $p \times p$ orthogonal matrices with $\eta'(\eta'\eta)^{-1/2}$ and $u(uu')^{-1/2}$ as their first rows, respectively. Then replacing C^* by $\tilde{C} = U_1 C^* U_2'$ and using the fact that $\tilde{q}(\cdot)$ satisfies (Exercise 5.2),

$$
\tilde{q}(RVR') = \tilde{q}(V) \qquad \text{for all} \quad V \in \bar{\mathscr{S}}(p), \quad R \in \mathcal{O}(p), \tag{5.53}
$$

we get

(5.54)

$$H(x\,|\,\Theta_1) = |\Sigma|^{(n_1+n_3)/2}\cdot c_1$$

$$\cdot\int_{\mathcal{O}(n_1)\times \mathrm{GL}(p)} \tilde{q}(U_1\tilde{C}u'u\tilde{C}'U_1' - U_1\tilde{C}u'Q_1'\eta'U_1' - U_1\eta Q_1u\tilde{C}'U_1'$$

$$+ U_1\eta'\eta U_1' + U_1\tilde{C}\tilde{C}'U_1')|\tilde{C}\tilde{C}'|^{k/2}\,d\tilde{C}\,dv(Q_1)$$

$$= |\Sigma|^{(n_1+n_3)/2}\cdot c_1$$

$$\cdot\int_{\mathcal{O}(n_1)\times \mathrm{GL}(p)} \tilde{q}(\tilde{C}U_2u'uU_2'\tilde{C}' - \tilde{C}U_2u'Q_1'\eta'U_1' - U_1\eta Q_1uU_2'\tilde{C}'$$

$$+ \delta\underset{\sim}{e}\underset{\sim}{e}' + \tilde{C}\tilde{C}')|\tilde{C}\tilde{C}'|^{k/2}\,d\tilde{C}\,dv(Q_1)$$

$$= |\Sigma|^{(n_1+n_3)/2}\cdot c_1$$

$$\cdot\int_{\mathcal{O}(n_1)\times \mathrm{GL}(p)} \tilde{q}\Bigg(T\underset{\sim}{a}_1\underset{\sim}{a}_1' - T^{1/2}\delta^{1/2}r_{11}(\underset{\sim}{a}_1\underset{\sim}{e}' + \underset{\sim}{e}\underset{\sim}{a}_1')$$

$$+ \delta\underset{\sim}{e}\underset{\sim}{e}' + \sum_1^p a_i a_i'\Bigg)|AA'|^{k/2}\,dA\,dv(Q_1),$$

where $T = \underset{\sim}{u}\underset{\sim}{u}' = X_1(X_3'X_3)^{-1}X_1'$ and $\delta = \eta'\eta = \mu_1\Sigma^{-1}\mu_1'$. Finally, transforming $\underset{\sim}{a}_1$ into $\underset{\sim}{a}_1/(1+T)^{1/2}$ and taking the ratio $H(x\,|\,\Theta_1)/H(x\,|\,\Theta_0)$ yields (5.49).

Next, suppose $p = 1$ and let V_1 and V_2 be $n_1 \times n_1$ orthogonal matrices with $\eta'(\eta\eta')^{-1/2}$ and $u(u'u)^{-1/2}$ as their first columns, respectively. Then replacing Q_1 in (5.52) by $V_1'Q_1V_2$ gives

(5.55)
$$H(x\,|\,\Theta_1) = |\Sigma|^{(n_1+n_3)/2}\cdot c_1$$

$$\cdot\int_{\mathcal{O}(n_1)\times \mathrm{GL}(p)} \tilde{q}(A^2T + A^2 + \delta - 2Ar_{11}\delta^{1/2}T^{1/2})$$

$$|AA'|^{k/2}\,dA\,dv(Q_1),$$

where $T = u'u = X_1'X_1/X_3'X_3$. Hence transforming A to $A/(1+T)^{1/2}$ yields (5.49). This completes the proof of the lemma.

We now refer to Assumption 3.3 in Section 3.5 of Chapter 3. It is easy to see that $K(v(x)\,|\,\Theta_1)$, which is the numerator of R in (5.49), apart from a constant, is symmetric and convex in v, its convexity following from that of $\tilde{q}$ (Exercise 5.3). Moreover, it follows from Corollary 3.4 in Section 3.3 of Chapter 3 that the null robustness of $\pi(x)$ holds in the class specified by (5.47). Hence the following result is deduced from Theorem 3.11 of Chapter 3.

Theorem 5.1. *Under the MANOVA model* (5.38) *with the error term* $\tilde{E}$ *distributed as in* (5.47) *and* q *satisfying* (5.46), *the test which rejects the MANOVA hypothesis* H *in* (5.39) *for large values of* T *is UMPI when* $\min(n_1, p) = 1$. *The test is also null-robust.*

By letting $n_1 = 1$ and $n_2 = 0$ in Theorem 5.1, we obtain

Corollary 5.1. *The Hotelling's* T^2*-test is UMPI for the problem* (5.39) *under the model* (5.38) *and with* $\tilde{E}$ *satisfying* (5.46), (5.47), *when* $n_1 = 1$ *and* $n_2 = 0$. *The test is null-robust.*

Letting $p = 1$ in Theorem 5.1 yields

Corollary 5.2. *The standard F-test in the ANOVA problem is UMPI in the class of elliptically symmetric convex densities.*

Remark 5.1. When $\min(n_1, p) = 1$, all four tests for the MANOVA hypothesis mentioned in Section 3.2 of Chapter 3 become equivalent and represent a robust UMPI test in the class specified in Theorem 5.1.

Remark 5.2. It is not difficult to show that the LRT of the MANOVA hypothesis (5.39) under the MANOVA model (5.38) with $q(\cdot)$nonincreasing in its argument rejects H for large values of $|I_{n_1} + X_1(X'_3X_3)^{-1}X'_1|$ (Exercise 5.4). The reader should note that this result does not depend on the assumption of the normality of $\tilde{E}$ but is rather general.

5.4. GMANOVA Problem

Let us recall from Section 5.2 the canonical form of the GMANOVA model

$$X \equiv \begin{array}{c} \begin{array}{ccc} p_1 & p_2 & p_3 \end{array} \\ \begin{pmatrix} X_{11} & X_{12} & X_{13} \\ X_{21} & X_{22} & X_{23} \\ X_{31} & X_{32} & X_{33} \end{pmatrix} \end{array} \begin{array}{c} n_1 \\ n_2 \\ n_3 \end{array} = \begin{pmatrix} \Theta_{11} & \Theta_{12} & 0 \\ \Theta_{21} & \Theta_{22} & 0 \\ 0 & 0 & 0 \end{pmatrix} + \tilde{E} \tag{5.56}$$

with

$$\tilde{E}: n \times p \sim \mathscr{F}_{\mathrm{E}}(0, I_n \otimes \Sigma), \Sigma \in \mathscr{S}(p) \quad \text{unknown} \tag{5.57}$$

and the GMANOVA hypothesis

$$\mathrm{H}: \Theta_{12} = 0 \quad \text{versus} \quad \mathrm{K}: \Theta_{12} \neq 0. \tag{5.58}$$

To analyse the problem via invariance, let

$$
(5.59) \quad \mathscr{A} = \left\{ \begin{pmatrix} A_{11} & A_{12} & A_{13} \\ 0 & A_{22} & A_{23} \\ 0 & 0 & A_{33} \end{pmatrix} \in \mathrm{GL}(p) \,\Big|\, A_{ii} \in \mathrm{GL}(p_i), \quad i = 1, 2, 3 \right\}
$$

$$
(5.60) \quad \mathscr{F} = \left\{ F : n \times p \,\Big|\, F = \begin{pmatrix} \overset{p_1}{F_{11}} & \overset{p_2}{0} & \overset{p_3}{0} \\ F_{21} & F_{22} & 0 \\ 0 & 0 & 0 \end{pmatrix} \begin{matrix} n_1 \\ n_2 \\ n_3 \end{matrix} \right\}
$$

and

$$
(5.61) \quad \mathscr{P} = \left\{ P = \begin{pmatrix} P_1 & 0 & 0 \\ 0 & P_2 & 0 \\ 0 & 0 & P_3 \end{pmatrix} \in \mathcal{O}(n) \,\Big|\, P_i \in \mathcal{O}(n_i), \quad i = 1, 2, 3 \right\}
$$

$$
= \mathcal{O}(n_1) \times \mathcal{O}(n_2) \times \mathcal{O}(n_3).
$$

Consider the group

$$
(5.62) \quad \mathscr{G} = \mathscr{P} \times \mathscr{A} \times \mathscr{F} = \mathcal{O}(n_1) \times \mathcal{O}(n_2) \times \mathcal{O}(n_3) \times \mathscr{A} \times \mathscr{F}
$$

with group operation

$$
(P_{(2)}, A_{(2)}, F_{(2)}) \cdot (P_{(1)}, A_{(1)}, F_{(1)}) = (P_{(2)} P_{(1)}, A_{(2)} A_{(1)}, P_{(2)} F_{(1)} A'_{(2)} + F_{(2)})
$$

for $(P_{(i)}, A_{(i)}, F_{(i)}) \in \mathscr{G}$, $i = 1, 2$. The left action of $\mathscr{G}$ on X is given by

$$
(5.63) \quad g(X) = P X A' + F \quad \text{for} \quad g = (P, A, F) \in \mathscr{G},
$$

and the group acting on the parameter space is $\bar{\mathscr{G}} = \mathscr{G}$ with the left action

$$
(5.64) \quad \bar{g}(\Theta, \Sigma) = (P \Theta A' + F, A \Sigma A') \quad \text{for} \quad \bar{g} = (P, A, F) \in \bar{\mathscr{G}}.
$$

It is easy to see that the distribution of $g(X)$ under $\bar{g}(\Theta, \Sigma)$ is the same as that of X under (Θ, Σ), and that $\bar{\mathscr{G}} = \mathscr{G}$ preserves the hypothesis. Hence the problem (5.58) is left invariant under $\mathscr{G}$. We shall call this group $\mathscr{G}$ the full group.

To describe the nature and distribution of $\pi(x)$, a maximal invariant under $\mathscr{G}$, it is more convenient to work with a smaller group, $\mathscr{H} = \mathcal{O}(n_3) \times \mathscr{A} \times \mathscr{F}$, which is isomorphic to the subgroup $\{I_{n_1}\} \times \{I_{n_2}\} \times \mathcal{O}(n_3) \times \mathscr{A} \times \mathscr{F}$ of the group $\mathscr{G}$. This brings out some salient features of the GMANOVA problem and also enables us to work with a subset of X as demonstrated below. Clearly $\mathscr{H}$ also leaves the testing problem (5.58) invariant under the left action

$$
(5.65) \quad h(X) = \tilde{P} X A' + F, \quad \bar{h}(\Theta, \Sigma) = (\tilde{P} \Theta A' + F, A \Sigma A'),
$$

where

$$h = \bar{h} = (P_3, A, F) \in \mathscr{H} \equiv \bar{\mathscr{H}} \text{ with } \tilde{P} = \text{diag}(I_{n_1}, I_{n_2}, P_3).$$

Here diag(A, B, C) denotes the block diagonal matrix with A, B, and C as diagonal blocks. Note that under the subgroup $\mathscr{O}(n_3)$ acting on X by $X \to \tilde{P}X$, the sufficient statistic $(\bar{X}, V)$ in (5.36) is a maximal invariant. This implies that a maximal invariant under $\mathscr{H}$ is a function of $(\bar{X}, V)$ and that under the full group $\mathscr{G}$, a maximal invariant is also a function of $(\bar{X}, V)$ since $\mathscr{G}$-invariance implies $\mathscr{H}$-invariance. More explicitly, the following propositions provide a maximal invariant under $\mathscr{H}$ and a maximal invariant parameter under $\bar{\mathscr{H}}$. The proofs are given in the Appendix.

Proposition 5.1. (*Gleser and Olkin*, 1970). *A maximal invariant under the group $\mathscr{H}$ is $s(X) = (s_1(X), s_2(X))$ where*

$$s_1(X) = (X_{12}, X_{13})\begin{pmatrix} V_{22} & V_{23} \\ V_{32} & V_{33} \end{pmatrix}^{-1}(X_{12}, X_{13})' \tag{5.66}$$

$$s_2(X) = \begin{pmatrix} X_{13} \\ X_{23} \end{pmatrix} V_{33}^{-1}(X'_{13}, X'_{23}).$$

In the above, $V \equiv (V_{ij}) = (X'_{3i}X_{3j})$: $p \times p$ as defined in (5.36).

Proposition 5.2. (*Gleser and Olkin*, 1970). *Under $\bar{\mathscr{H}} = \mathscr{H}$, a maximal invariant parameter is $\Theta_{12}\Sigma_{22.3}^{-1}\Theta'_{12} \equiv \xi\xi'$ with $\xi = \Theta_{12}\Sigma_{22.3}^{-1/2}$, where $\Sigma_{22.3} = \Sigma_{22} - \Sigma_{23}\Sigma_{33}^{-1}\Sigma_{32}$.*

Hence, when considering the group $\mathscr{G}$ and its maximal invariant $\pi(x)$, we can restrict our attention to

$$\bar{\bar{X}} \equiv (U_1, U_2, U_3) \equiv ((X_{12}, X_{13}), X_{23}, (X_{32}, X_{33})), \tag{5.67}$$

and without loss of generality set $\Theta_{11} = 0$, $\Theta_{21} = 0$, $\Theta_{22} = 0$, $\Sigma = I_p$, and replace Θ_{12} by ξ. From (5.56)–(5.57), the marginal probability density function of $\bar{\bar{X}}$ is obtained as

$$\bar{f}(\bar{\bar{X}} \mid \xi) = \bar{q}(\text{tr}(U_1 - \xi^*)'(U_1 - \xi^*) + \text{tr}\, U'_2U_2 + \text{tr}\, U'_3U_3), \tag{5.68}$$

where $\xi^* = (\xi, 0)$: $n_1 \times (p_2 + p_3)$, and

$$\bar{q}(u) = \int_L q(u + \text{tr}\, X'_{11}X_{11} + \text{tr}\, X'_{21}X_{21} + \text{tr}\, X'_{31}X_{31} + \text{tr}\, X'_{22}X_{22})dX_{11}dX_{21}dX_{31}dX_{22}, \tag{5.69}$$

where $L = R^{n_1p_1} \times R^{n_2p_1} \times R^{n_3p_1} \times R^{n_2p_2}$.

In view of the restriction to $\bar{\bar{X}}$, the group $\mathscr{G}$ acting on the left of X is also reduced to the subgroup $\bar{\bar{\mathscr{G}}} = \mathscr{O}(n_1) \times \bar{\mathscr{A}}$ acting on the left of $\bar{\bar{X}}$ as

(5.70)
$$\bar{\bar{g}}(\bar{\bar{X}}) = \bar{\bar{g}}(U_1, U_2, U_3) = (P_1 U_1 \bar{A}', X_{23}A'_{33}, U_3\bar{A}') \quad \text{for} \quad \bar{\bar{g}} = (P_1, \bar{A}) \in \bar{\bar{\mathscr{G}}},$$

when $\bar{\mathscr{A}}$ is the group of $(p_2 + p_3) \times (p_2 + p_3)$ nonsingular matrices of the form $\bar{A} = (A_{ij})$, $i, j = 2, 3$ with $A_{32} = 0$. Here the subgroup $\mathscr{O}(n_2) \times \mathscr{O}(n_3)$ of $\mathscr{G}$ is ignored, since it does not affect U_2, U_3 in (5.67).

We are now in a position to evaluate the probability ratio R of nonnull to null distribution of $\pi(x)$ based on (5.68) and (5.69). See (3.47) and (3.48) in Section 3.4 of Chapter 3, providing an expression for R. The numerator of R, namely $H(x \mid \theta_1)$, is given by

(5.71)
$$\int_{\mathscr{G}} \bar{q}(\operatorname{tr}(P_1U_1\bar{A}' - \xi^*)'(P_1U_1\bar{A}' - \xi^*) + \operatorname{tr} A_{33}X'_{23}X_{23}A'_{33} + \operatorname{tr} \bar{A}U'_3U_3\bar{A}')\mu(d\bar{A})\nu(dP_1)$$

where $\mu(d\bar{A}) = (|A_{22}A'_{22}|^{(M-p_2)/2}\, dA_{22})(|A_{33}A'_{33}|^{(n-p_3)/2}\, dA_{33})\, dA_{23}$ with $M = n_1 + n_3 - p_3$, and ν is the invariant probability measure on $\mathscr{O}(n_1)$. It is noted that a left-invariant measure on $\bar{\mathscr{A}}$ is $(|A_{22}A'_{22}|^{-(p_2+p_3)/2}\, dA_{22})$ $(|A_{33}A'_{33}|^{-p_3/2}\, dA_{33})\, dA_{23}$ and that the inverse of the Jacobian of transformation (5.70) is $|A_{22}A'_{22}|^{(n_1+n_3)/2}|A_{33}A'_{33}|^{n/2}$. Here transforming $\bar{A}$ into $\bar{A}C$ in (5.71), where $C = (C_{ij}) \in \bar{\mathscr{A}}$ and $C(U'_1U_1 + U'_3U_3)C' = I_{p_2+p_3}$, the inside of $\bar{q}$ in (5.71) becomes

(5.72)
$$\operatorname{tr} \bar{A}\bar{A}' - 2\operatorname{tr} \xi' P_1(W_2A'_{22} + W_3A'_{23}) + \operatorname{tr} A_{33}X'_{23}X_{23}A'_{33} + \delta,$$

where $\delta = \operatorname{tr} \xi\xi' = \operatorname{tr} \Theta_{12}\Sigma^{-1}_{22.3}\Theta'_{12}$, and

(5.73)
$$(W_2, W_3) \equiv U_1C' = (X_{12}C'_{22} + X_{13}C'_{23}, X_{13}C'_{33}).$$

We always ignore multiplicative constants coming out by transformations when they are cancelled out with those of the denominator $H(x \mid \theta_0)$. Further, transforming A_{33} into $A_{33}(I + X'_{23}X_{23})^{-1/2}$, the inside of $\bar{q}$ or (5.72) becomes, for example,

(5.74)
$$\operatorname{tr} \bar{A}\bar{A}' - 2\operatorname{tr} \xi' P_1(W_2A'_{22} + W_3A'_{23}) + \delta \equiv \operatorname{tr} \bar{A}\bar{A}' - 2\eta + \delta.$$

Under these transformations, the ratio becomes (writing ξ in place of θ_1 and 0 in place of θ_0)

(5.75)
$$R \equiv dP^{\pi}_{\xi}/dP^{\pi}_{0} = \int_{\mathscr{G}} \bar{q}(\operatorname{tr} \bar{A}\bar{A}' - 2\eta + \delta)\nu(dP_1)\mu(d\bar{A}) \Big/ \int_{\mathscr{G}} \bar{q}(\operatorname{tr} \bar{A}\bar{A}')\nu(dP_1)\mu(d\bar{A}).$$

Here since R is pdf of π with respect to P_0^{π}, and since C in (5.73) does not depend on X_{23}, so that R does not depend on X_{23}, it follows that the density of π evaluated at $\pi(U_1, U_2, U_3)$ does not involve U_2. This implies that the density is a function of π_0, where π_0 is a maximal invariant function of (U_1, U_3) under the action (5.70) with U_2 part ignored. Therefore, since P_0^{π} is independent of any parameter, π_0 is sufficient for the family $\{P_{\xi}^{\pi} \mid \xi \in R^{n_1 \times p_2}\}$ by Neyman's factorization theorem. This result also follows from the Stein theorem (Hall et al., 1965) as in the normal case (see Kariya, 1978, Theorem 3.1). Thus we obtain

Theorem 5.2. *The class of invariant tests based on* $s_1(X)$ *in* (5.66) *and* $\pi_2 \equiv X_{13}V_{33}^{-1}X'_{13}$ *only forms an essentially complete class.*

Several interesting points pertaining to this result are worth mentioning. First, we note that in this theorem no assumption is made on q or $\bar{q}$ in (5.69). Second, this theorem implies that the statistics $\pi_3 \equiv X_{13}V_{33}^{-1}X'_{23}$ and $\pi_4 \equiv X_{23}V_{33}^{-1}X'_{23}$ in the maximal invariant (5.66) under $\mathscr{H} = \mathscr{O}(n_3) \times \mathscr{A} \times \mathscr{F}$ can be discarded when an invariant test under $\mathscr{G}$ is considered. Third, this result can also be derived via the distributional properties of $s_1(X)$ and $s_2(X)$, especially under the normality of $\tilde{E}$ (see Kariya, 1978). However, the above argument holds for any q, gives a simpler proof for the normal case, and makes it clear that the result is more related to the invariant structure of the problem rather than to the distributional structure. Of course, when $p_3 = 0$, or especially in the MANOVA problem, the result in Theorem 5.2 becomes trivial.

Next, following the general procedure outlined in Section 3.5 of Chapter 3, we need to expand the ratio R in (5.75) around $\xi = 0$. For this we assume

Assumption 5.1. The function q in the pdf in (5.57) belongs to $\mathscr{Q}$, where $\mathscr{Q}$ is the class of three times continuously differentiable functions from $[0, \infty)$ into $[0, \infty)$ such that

(5.76) $$\int_{R^{np}} q(\operatorname{tr} Z'Z)\, dZ = 1,$$

(5.77) $$\int_{\bar{\mathscr{A}}} (\operatorname{tr} \bar{A}\bar{A}')^{i/2} \, |\bar{q}^{(i)}(\operatorname{tr} \bar{A}\bar{A}')| \, \mu(d\bar{A}) < \infty \qquad (i = 1, 2, 3),$$

(5.78) $$\bar{q}^{(3)}(x) \leqq 0, \qquad \text{and} \quad \bar{q}^{(3)} \quad \text{is nondecreasing},$$

where $\bar{q}^{(i)}(x) = d^i\bar{q}(x)/dx^i$.

Under this assumption, we prove the following result.

Theorem 5.3. (*GMANOVA*). *Under Assumption* 5.1, *the test based on the critical region*

$$a_0 \operatorname{tr} Z(Z'Z + V_{22.3})^{-1}Z(I + \pi_2)^{-1} - \operatorname{tr}(I + \pi_2)^{-1} > c \tag{5.79}$$

with

$$a_0 = (n_1 + n_3 - p_3)/2, \qquad Z = (I + \pi_2)^{-1/2}(X_{12} - X_{13}V_{33}^{-1}V_{32}) \tag{5.80}$$

and

$$\pi_2 = X_{13}V_{33}^{-1}X'_{13}$$

is LBI for testing H: $\Theta_{12} = 0$ *versus* K: $\Theta_{12} \neq 0$ *under the distribution* (5.57).

Corollary 5.3. (*MANOVA*). *Under Assumption* 5.1 *and when* $p_1 = p_3 = 0$, *the test based on*

$$\operatorname{tr} X_{12}(X'_{12}X_{12} + V_{22})^{-1}X'_{12} > c \tag{5.81}$$

is LBI for testing H: $\Theta_{12} = 0$ *versus* K: $\Theta_{12} \neq 0$ *under the distribution* (5.57).

Corollary 5.3 follows directly from Theorem 5.3 by setting $p_1 = p_3 = 0$ formally. Before we prove the theorem, let us make two remarks: First, the LBI test in (5.79) under the distribution (5.57) is the same as the LBI test derived under normality. This is because the normal density satisfies Assumption 5.1 (Exercise 5.5). This implies that the LBI property is robust at least within the class of pdf's specified by Assumption 5.1. Similarly the LBI test in (5.81) coincides with the Pillai test in the MANOVA (see Section 3.2 of Chapter 3), the LBI property of which under normality is shown by Schwartz (1967a). Hence Corollary 5.3 also shows the robustness of the LBI property of the test up to the class $\mathscr{Q}$. Second, the conditions in Assumption 5.1 are satisfied for large class of pdf's, especially in the case of the normal scale mixture: $q(x) = \int_0^\infty e^{-ax/2}a^{np/2}(2\pi)^{-np/2}\,dF(a)$, provided some conditions on the moments hold (Exercise 5.6).

Proof of Theorem 5.3. Expand the integrand of $\bar{q}$ in the numerator of (5.75) as

$$\bar{q}(\operatorname{tr}\bar{A}\bar{A}') + \bar{q}^{(1)}(\operatorname{tr}AA')(-2\eta + \delta) + \tfrac{1}{2}\bar{q}^{(2)}(\operatorname{tr}\bar{A}\bar{A}')(-2\eta + \delta)^2 + \tfrac{1}{6}\bar{q}^{(3)}(z)(-2\eta + \delta)^3, \tag{5.82}$$

where $z = \operatorname{tr}\bar{A}\bar{A}' + (1 - \alpha)(-2\eta + \delta)$ with $0 \leqq \alpha \leqq 1$. We evaluate the integration of each term in (5.82). First, since the integral of $(\operatorname{tr} P_1 Q)^k$ over $\mathscr{O}(n_1)$ with respect to $\nu(dP_1)$ is zero for k odd (see Chapter 2), from (5.74) and (5.75) the integration of the second term of (5.82) over $\bar{\bar{\mathscr{G}}}$ is simply $\delta \int_{\bar{\mathscr{A}}} \bar{q}^{(1)}(\operatorname{tr}\bar{A}\bar{A}')\mu(d\bar{A})$.

Second, the integration of the third term of (5.82) becomes

$$\text{(5.83)}\quad \begin{aligned}&\tfrac{2}{n_1}\int_{\bar{\mathscr{A}}}[\operatorname{tr}(W_2A'_{22}+W_3A'_{23})\xi'\xi(W_2A'_{22}+W_3A'_{23})']\bar{q}^{(2)}(\operatorname{tr}\bar{A}\bar{A}')\mu(d\bar{A})\\ &+\tfrac{\delta^2}{2}\int_{\bar{\mathscr{A}}}\bar{q}^{(2)}(\operatorname{tr}\bar{A}\bar{A}')\mu(d\bar{A}),\end{aligned}$$

since $\int_{\mathscr{O}(n_1)}(\operatorname{tr} P_1Q)^2\nu(dP_1)=\operatorname{tr} Q'Q/n_1$ (see Lemma 2.5 in Chapter 2). In the first term of (5.83), the measure $\bar{q}^{(2)}(\operatorname{tr}\bar{A}\bar{A}')\mu(d\bar{A})$ is invariant under the sign change $A_{22}\to -A_{22}$ and so the integration of $\operatorname{tr} W_2A'_{22}\xi'\xi A_{23}W'_3$ is zero. To evaluate the integration of $\operatorname{tr} W_2A'_{22}\xi'\xi A_{22}W'_2$ in (5.83), let $\bar{\bar{q}}^{(2)}(\operatorname{tr} A_{22}A'_{22})\mu_2(dA_{22})$ be the marginal measure of A_{22} where $\mu_2(dA_{22})=|A_{22}A'_{22}|^{(M-p_2)/2}\,dA_{22}$ and decompose $GL(p_2)=G_T(p_2)\times\mathscr{O}(p_2)$ and $\mu_2(dA_{22})=\lambda_2(dB_2)\tau_2(dQ_2)$, where $G_T(p_2)$ is the group of $p_2\times p_2$ lower triangular matrices with positive diagonal elements, $\tau_2(dQ_2)$ is the invariant probability meausre on $\mathscr{O}(p_2)$, and $\lambda_2(dB_2)=|B_2B'_2|^{M/2}(\Pi b_{ii}^{-i})\,dB_2$ with $B_2=(b_{ij})$ (see, e.g., Wijsman, 1967, p. 398 or Eaton, 1983, p. 213). Note that $(\Pi b_{ii}^{-i})\,dB_2$ is a left-invariant measure on $G_T(p_2)$. Under this decomposition, with $A_{22}=B_2Q_2$, the integration of $\operatorname{tr} W_2A'_{22}\xi'\xi A_{22}W'_2$, after integration over $\mathscr{O}(p_2)$, is

$$\text{(5.84)}\quad \frac{1}{p_2}(\operatorname{tr} W'_2W_2)\int_{G_T(p_2)}(\operatorname{tr} B_2B'_2\xi'\xi)\bar{\bar{q}}^{(2)}(\operatorname{tr} B_2B'_2)\lambda_2(dB_2)=\frac{\delta\beta_2}{p_2}\operatorname{tr} W'_2W_2$$

with

$$\text{(5.85)}\quad \beta_2p_2=\int_{G_T(p_2)}(\operatorname{tr} B_2B'_2)\bar{\bar{q}}^{(2)}(\operatorname{tr} B_2B'_2)\lambda_2(dB_2),$$

where

$$\int_{\mathscr{O}(p_2)}\operatorname{tr} AQ_2BQ'_2\tau_2(dQ_2)=\operatorname{tr} A\operatorname{tr} B/p_2$$

and

$$\int_{G_T(p_2)}B_2B'_2\bar{\bar{q}}^{(2)}(\operatorname{tr} B_2B'_2)\lambda_2(dB_2)=\beta_2I$$

are used. The first result appears in Lemma 2.5 of Chapter 2, and the second result follows from the fact that the integral of the off-diagonal elements of $B_2B'_2$ is zero (odd function) and that of the diagonal elements is the same (by symmetry). Further, to evaluate the integration of $\operatorname{tr} W_3A'_{23}\xi'\xi A_{23}W'_3$ in (5.83), let $\bar{\bar{q}}^{(2)}(\operatorname{tr} A_{23}A'_{23})\,dA_{23}$ be the marginal measure of A_{23}. Then a direct evaluation of the integration of $\operatorname{tr} W_3A'_{23}\xi'\xi A_{23}W'_3$ with respect to this measure yields $\delta\beta_3 \operatorname{tr} W_3' W_3$ with

$$(5.86)\qquad \beta_3 p_2 p_3 = \int_{R^{p_2 p_3}} (\operatorname{tr} A_{23}A_{23}')\tilde{\bar{q}}^{(2)}(\operatorname{tr} A_{23}A_{23}')\, dA_{23}.$$

Therefore, (5.83) is finally evaluated as

$$(5.87)\qquad (2/n_1)\delta[(\beta_2/p_2)\operatorname{tr} W_2' W_2 + \beta_3 \operatorname{tr} W_3' W_3] + o(\delta).$$

Third, we show that the integration of the fourth term in (5.82) is $o(\delta)$. Since $|\eta| \leqq (\operatorname{tr} \bar{A}\bar{A}')^{1/2}(\operatorname{tr} \xi' HH'\xi)^{1/2} \leqq (\operatorname{tr} \bar{A}\bar{A}')^{1/2}\delta^{1/2}$ from $W_2W_2' + W_3W_3' \leqq I$ where $H = (P_1W_2, P_1W_3)$, $z \geqq \operatorname{tr} \bar{A}\bar{A}' - 2(\operatorname{tr} \bar{A}\bar{A}')^{1/2}\delta^{1/2} + \delta$. Hence, from (5.78),

(5.88)

$$\left|\int_{\bar{\bar{\mathscr{G}}}} \bar{q}^{(3)}(z)(-2\eta + \delta)^3 v(dP_1)\mu(d\bar{A})\right| \leqq \int_{\mathscr{G}} |\bar{q}^{(3)}(z)||-2\eta + \delta|^3 v(dP_1)\mu(d\bar{A})$$

$$\leqq \int_{\bar{\bar{\mathscr{G}}}} |\bar{q}^{(3)}(\operatorname{tr} \bar{A}\bar{A}' - 2(\operatorname{tr} \bar{A}\bar{A}')^{1/2}\delta^{1/2} + \delta)||-2\eta + \delta|^3 v(dP_1)\mu(d\bar{A}).$$

Here, since

$$|-2\eta + \delta|^3 \leqq 8(\operatorname{tr} \bar{A}\bar{A}')^{3/2}\delta^{3/2} + 12(\operatorname{tr} \bar{A}\bar{A}')\delta^2 + 6(\operatorname{tr} \bar{A}\bar{A}')^{1/2}\delta^{5/2} + \delta^3,$$

and since $\delta \geqq 0$, the right side of (5.88) is further bounded above by

$$(5.89)\qquad \sum_{i=1}^{4} c_i \int_{\mathscr{A}} |\bar{q}^{(3)}(\operatorname{tr} \bar{A}\bar{A}' - 2(\operatorname{tr} \bar{A}\bar{A}')^{1/2}\delta^{1/2})|(\operatorname{tr} \bar{A}\bar{A}')^{(4-i)/2}\delta^{(1+i)/2}\mu(d\bar{A}),$$

where $c_1 = 8, c_2 = 12, c_3 = 6$, and $c_4 = 1$. We split the domain of this integral into $E_0 = \{\operatorname{tr} \bar{A}\bar{A}' \leqq 1\}$ and $E_1 = \{\operatorname{tr} \bar{A}\bar{A}' > 1\}$. On E_1, by (5.78) replacing $(\operatorname{tr} \bar{A}\bar{A}')^{1/2}$ in the inside of $\bar{q}^{(3)}$ by $\operatorname{tr} \bar{A}\bar{A}'$, and changing $\bar{A}$ into $(1 - 2\delta^{1/2})^{1/2}\bar{A}$ for $\delta < 1/8$, (5.89) is shown to be $o(\delta)$, while on E_0, it is $o(\delta)$ from the boundedness of $\bar{q}^{(3)}$.

Consequently, noticing that the denominator of the ratio R in (5.75) is simply $H(x\,|\,0) = \int_{\bar{\mathscr{A}}} \bar{q}(\operatorname{tr} \bar{A}\bar{A}')\mu(d\bar{A}) \equiv D$, we obtain

Lemma 5.3. *The ratio R in (5.75) is evaluated as*

$$R = 1 + \frac{\delta}{D}\left[\beta_1 + \frac{2}{n_1}\left(\frac{\beta_2}{p_2}\operatorname{tr} W_2'W_2 + \beta_3 \operatorname{tr} W_3'W_3\right)\right] + o(\delta),$$

where $\beta_1 = \int_{\bar{\mathscr{A}}} \bar{q}^{(1)}(\operatorname{tr} \bar{A}\bar{A}')\mu(d\bar{A})$, the W_i's are defined in (5.73), β_2 and β_3 are given by (5.85) and (5.86), respectively, and $o(\delta)$ is uniform in Z.

This is an expression of (3.94) in Section 3.5 of Chapter 3 as applied to the GMANOVA problem in the canonical form.

The quantities $\operatorname{tr} W_2' W_2$, $\operatorname{tr} W_3' W_3$, and β_2/β_3 are identified below.

Lemma 5.4.

(1) $\operatorname{tr} W_2' W_2 = \operatorname{tr} Z(Z'Z + V_{22.3})^{-1} Z'(I + \pi_2)^{-1}$,

(2) $\operatorname{tr} W_3' W_3 = -\operatorname{tr}(I + \pi_2)^{-1} + n_1$,

(3) $\beta_2/\beta_3 = M \equiv n_1 + n_3 - p_3$.

Proof. The proof of (1) and (2) is straightforward and given in the Appendix. To prove (3), consider the marginal density of A_{22} and A_{23} with respect to $|A_{22}A_{22}'|^{(M-p_2)/2}\, dA_{22}\, dA_{23}$:

$$h(\operatorname{tr} A_{22}A_{22}' + \operatorname{tr} A_{23}A_{23}') = \int_{\mathrm{GL}(p_3)} \bar{q}^{(2)}(\operatorname{tr} A_{22}A_{22}' + \operatorname{tr} A_{23}A_{23}' + \operatorname{tr} A_{33}A_{33}') \times |A_{33}A_{33}'|^{(n-p_3)/2}\, dA_{33},$$

and as we did before, decompose A_{22} as $A_{22} = B_2 Q_2$ with $Q_2 \in \mathcal{O}(p_2)$ and $B_2 \in G_{\mathrm{T}}(p_2)$. Then the marginal measure of B_2 and A_{23} is given by

(5.90)
$$\psi(dB_2, dA_{23}) \equiv h(\operatorname{tr} B_2 B_2' + \operatorname{tr} A_{23}A_{23}')|B_2 B_2'|^{M/2}(\Pi(b_{ii}^2)^{-i/2})\, dB_2\, dA_{23}.$$

Note $\beta_2 p_2 = \int \operatorname{tr} B_2 B_2' \psi(dB_2, dA_{23})$ and $\beta_3 p_2 p_3 = \int \operatorname{tr} A_{23}A_{23}' \psi(dB_2, dA_{23})$. Define $B_2 = (b_{ij})$, $A_{23} = (a_{ij})$, $L = \operatorname{tr} B_2 B_2' + \operatorname{tr} A_{23}A_{23}'$, $e_0 = \operatorname{tr} A_{23}A_{23}'/L = \Sigma a_{ij}^2/L$, $e_i = b_{ii}^2/L$ $(i = 1, \ldots, p_2)$, $e_{i+p_2} = b_{i+1,i}^2/L$ $(i = 2, \ldots, p_2 - 1)$, $e_{i+p_2+(p_2-1)} = b_{i+2,i}^2/L$ $(i = 3, \ldots, p_2 - 2), \ldots,$ and $e_{p_2(p_2+1)/2}/L = b_{p_2,1}^2$. Further extend the domain $[0, \infty)$ of b_{ii} into $(-\infty, \infty)$, which simply gives a multiplicative constant, say c_0, to the right side of (5.90) and let $K = \int c_0 h(\operatorname{tr} B_2 B_2' + \operatorname{tr} A_{23}A_{23}')\, dB_2\, dA_{23}$. Then, since $c_0 h(\operatorname{tr} B_2 B_2' + \operatorname{tr} A_{23}A_{23}')/K$ is a spherical density of b_{ij}'s and a_{ij}'s, L and $e \equiv (e_0, e_1, \ldots, e_{p_2(p_2+1)/2})$ are independent and e obeys the Dirichelet distribution $D(p_1 p_2/2,\ 1/2, \ldots, 1/2)$ (see, e.g., Kariya and Eaton, 1977). Hence, letting $N = \int L^{\Sigma(M-i)/2} c_0 \times h(\operatorname{tr} B_2 B_2' + \operatorname{tr} A_{23}A_{23}')\, dB_2\, dA_{23}$, which is finite from $\beta_2 < \infty$ and $\beta_3 < \infty$, we obtain

$$\beta_3 p_2 p_3/KN = E\left(e_0 \prod_{i=1}^{p_2} e_i^{(M-i)/2}\right).$$

This is directly evaluated from the formula (77.9) of Wilks (1962) on page 179. Similarly, evaluating

$$\beta_2 p_2/KN = E\left[\sum_{j=1}^{p_2(p_2+1)} e_j \prod_{i=1}^{p_2} e_i^{(M-i)/2}\right] = \sum_{i=1}^{p_2} E\left(\prod_{i=1}^{p_2} e_i^{(M-i)/2}\right) + \sum_{j=p_2+1}^{p_2(p_2+1)/2} E\left(e_j \prod_{i=1}^{p_2} e_i^{(M-i)/2}\right)$$

and computing the ratio $\beta_2 p_2/\beta_3 p_2 p_3$ yields

$$\sum_{i=1}^{p_2} (M-i+1)/p_2 p_3 + \sum_{i>j} 1/p_2 p_3 = M/p_3.$$

Hence $\beta_2/\beta_3 = M$, completing the proof of the lemma.

By using Lemma 5.3 and Lemma 5.4, the proof of the theorem is complete.

Remark 5.3. It is not difficult to derive the LRT of the GMANOVA hypothesis (5.58) under the GMANOVA model (5.56) when $q(\cdot)$ in $\mathscr{F}_E$ is nonincreasing in its argument. Denoting the LRT test statistic by $\lambda \equiv \sup_{\Theta_0} f(x\,|\,\theta)/\sup_{\Theta} f(x\,|\,\theta)$, it follows that (Exercise 5.7)

$$\lambda^{-2/n} = D_1/D_0 \tag{5.91}$$

where

$$D_1 = \begin{vmatrix} X'_{31}X_{31} & X'_{31}X_{32} & X'_{31}X_{33} \\ X'_{32}X_{31} & X'_{12}X_{12} + X'_{32}X_{32} & X'_{12}X_{13} + X'_{32}X_{33} \\ X'_{33}X_{31} & X'_{13}X_{12} + X'_{33}X_{32} & X'_{13}X_{13} + X'_{23}X_{23} + X'_{33}X_{33} \end{vmatrix}$$

and

$$D_0 = \begin{vmatrix} X'_{31}X_{31} & X'_{31}X_{32} & X'_{31}X_{33} \\ X'_{32}X_{31} & X'_{32}X_{32} & X'_{32}X_{33} \\ X'_{33}X_{31} & X'_{33}X_{32} & X'_{13}X_{13} + X'_{23}X_{23} + X'_{33}X_{33} \end{vmatrix}.$$

Using $\begin{vmatrix} A & B \\ B' & C \end{vmatrix} = |C||A - BC^{-1}B'|$, the ratio D_1/D_0 simplifies to D_1^*/D_0^*, where

$$D_1^* = \left| \begin{pmatrix} X'_{31}X_{31} & X'_{31}X_{32} \\ X'_{32}X_{31} & X'_{12}X_{12} + X'_{32}X_{32} \end{pmatrix} - \begin{pmatrix} X'_{13}X_{33} \\ X'_{12}X_{13} + X'_{32}X_{33} \end{pmatrix} \times \left(\sum_1^3 X'_{i3}X_{i3}\right)^{-1} \begin{pmatrix} X'_{13}X_{33} \\ X'_{12}X_{13} + X'_{32}X_{33} \end{pmatrix}' \right| \tag{5.92}$$

and

$$D_0^* = \left|\begin{pmatrix} X'_{31}X_{31} & X'_{31}X_{32} \\ X'_{32}X_{31} & X'_{32}X_{32} \end{pmatrix} - \begin{pmatrix} X'_{13}X_{33} \\ X'_{32}X_{33} \end{pmatrix}\left(\sum_1^3 X'_{i3}X_{i3}\right)^{-1}\begin{pmatrix} X'_{13}X_{33} \\ X'_{32}X_{33} \end{pmatrix}'\right|.$$

But the matrix C_1 (say) within D_1^* can be written as

(5.93)

$$C_1 = C_0 + \left[\begin{pmatrix} 0 & 0 \\ 0 & X'_{12}X_{12} \end{pmatrix} - \begin{pmatrix} 0 & X'_{13}X_{33}\left(\sum_1^3 X'_{i3}X_{i3}\right)^{-1} X'_{13}X_{12} \\ X'_{12}X_{13}\left(\sum_1^3 X'_{i3}X_{i3}\right)^{-1} X'_{33}X_{13} & \Lambda \end{pmatrix}\right]$$

where C_0 is the matrix within D_0^* and

(5.94)
$$\Lambda = X'_{12}X_{13}\left(\sum_1^3 X'_{i3}X_{i3}\right)^{-1} X'_{13}X_{12} + X'_{12}X_{13}\left(\sum_1^3 X'_{i3}X_{i3}\right)^{-1} X'_{33}X_{32} + X'_{32}X_{33}\left(\sum_1^3 X'_{i3}X_{i3}\right)^{-1} X'_{13}X_{12}.$$

It then follows that

(5.95)
$$\lambda^{-2/n} = |I_{n_1} + C_0^{-1/2}(C_1 - C_0)C_0^{-1/2}|.$$

The reader is invited to verify that (Exercise 5.8)

(5.96)
$$C_0^{-1/2}(C_1 - C_0)C_0^{-1/2} = ZV_{22.3}^{-1}Z'$$

where $Z = (I + \pi_2)^{-1/2}(X_{12} - X_{13}V_{33}^{-1}V_{32})$, and $\pi_2 = X_{13}V_{33}^{-1}X'_{13}$ as defined in Theorem 5.3.

Remark 5.4. Based on the matrix $ZV_{22.3}^{-1}Z'$, Khatri (1966) proposed the following tests for the GMANOVA problem that are in the same spirit as in the MANOVA problem (see Section 3.2 of Chapter 3).

(5.97) Roy's maximum root test based on $Ch_1(ZV_{22.3}^{-1}Z')$;

(5.98) Lawley–Hotelling's trace test based on $\operatorname{tr} ZV_{22.3}^{-1}Z'$;

(5.99) Pillai's trace test based on $\operatorname{tr}(ZV_{22.3}^{-1}Z')(I_{n_1} + ZV_{22.3}^{-1}Z')^{-1}$.

Of course, these tests were proposed under the normality of $\tilde{E}$. However, although all of the above four tests are $\mathscr{G}$-invariant, none of them is LBI, and each of them is dominated locally by the LBI test derived in Theorem 5.3.

Appendix

Proof of Propositions 5.1 and 5.2. Clearly $s(X)$ is invariant, i.e., $s(X) = s(h(X))$ for $h \in \mathscr{H}$. To show its maximality, suppose $s(X) = s(X^*)$, i.e., from (5.66),

(A.5.1)

$$(X_{12}, X_{13})\begin{pmatrix} V_{22} & V_{23} \\ V_{32} & V_{33} \end{pmatrix}^{-1}(X_{12}, X_{13})' = (X_{12}^*, X_{13}^*)\begin{pmatrix} V_{22}^* & V_{23}^* \\ V_{32}^* & V_{33}^* \end{pmatrix}^{-1}(X_{12}^*, X_{13}^*)'$$

and

$$\text{(A.5.2)} \qquad \begin{pmatrix} X_{13} \\ X_{23} \end{pmatrix} V_{33}^{-1}(X'_{13}, X'_{23}) = \begin{pmatrix} X_{13}^* \\ X_{23}^* \end{pmatrix} V_{33}^{*-1}(X_{13}^{*\prime}, X_{23}^{*\prime}).$$

Let $V = SS'$ where $S \in \mathscr{A}$ defined in (5.59). Then

$$\text{(A.5.3)} \qquad \begin{aligned} \begin{pmatrix} V_{22} & V_{23} \\ V_{32} & V_{33} \end{pmatrix} &= \begin{pmatrix} S_{22} & S_{23} \\ 0 & S_{33} \end{pmatrix}\begin{pmatrix} S_{22} & S_{23} \\ 0 & S_{33} \end{pmatrix}' \\ &= \begin{pmatrix} S_{22}S'_{22} + S_{23}S'_{23} & S_{23}S'_{33} \\ S_{33}S'_{23} & S_{33}S'_{33} \end{pmatrix}. \end{aligned}$$

Similarly, let $V^* = S^*S^{*\prime}$ with $S^* \in \mathscr{A}$. This means

$$\text{(A.5.4)} \qquad \begin{pmatrix} V_{22}^* & V_{23}^* \\ V_{32}^* & V_{33}^* \end{pmatrix} = \begin{pmatrix} S_{22}^*S_{22}^{*\prime} + S_{23}^*S_{23}^{*\prime} & S_{23}^*S_{33}^{*\prime} \\ S_{33}^*S_{23}^{*\prime} & S_{33}^*S_{33}^{*\prime} \end{pmatrix}.$$

From (A.5.2) with $V_{33} = S_{33}S'_{33}$, for some $P_{33} \in \mathscr{O}(p_3)$,

$$\text{(A.5.5)} \qquad \begin{pmatrix} X_{13} \\ X_{23} \end{pmatrix} S_{33}^{\prime -1}P_{33} = \begin{pmatrix} X_{13}^* \\ X_{23}^* \end{pmatrix}(S_{33}^{*\prime})^{-1}.$$

Let $A'_{33} = S_{33}^{\prime -1}P_{33}S_{33}^{*\prime}$. Then from (A.5.4) and (A.5.5), we obtain

$$\text{(A.5.6)} \qquad A_{33}V_{33}A'_{33} = V_{33}^* \qquad \text{and} \qquad \begin{pmatrix} X_{13} \\ X_{23} \end{pmatrix} A'_{33} = \begin{pmatrix} X_{13}^* \\ X_{23}^* \end{pmatrix}.$$

Again, using the above relations, (A.5.1) is written as

$$(X_{12}S_{22}'^{-1} - X_{13}S_{33}'^{-1}S_{23}'S_{22}'^{-1})(X_{12}S_{22}'^{-1} - X_{13}S_{33}'^{-1}S_{23}'S_{22}'^{-1})'$$

(A.5.7) $$= (X_{12}^{*}S_{22}^{*\prime -1} - X_{13}^{*}S_{33}^{*\prime -1}S_{23}^{*\prime}S_{22}^{*\prime -1})$$

$$(X_{12}^{*}S_{22}^{*\prime -1} - X_{13}^{*}S_{33}'^{-1}S_{23}^{*\prime}S_{22}^{*\prime -1})'.$$

Hence for some $P_{22} \in \mathcal{O}(p_2)$, it holds that

(A.5.8) $$(X_{12}S_{22}'^{-1} - X_{13}S_{33}'^{-1}S_{23}'S_{22}'^{-1})P_{22} = X_{12}^{*}S_{22}^{*\prime -1} - X_{13}^{*}S_{33}'^{-1}S_{23}^{*\prime}S_{22}^{*\prime -1}.$$

From (A.5.6), (A.5.8), and the definition of A_{33}, we get

(A.59)

$$X_{12}S_{22}'^{-1}P_{22}S_{32}^{*\prime} + X_{13}[A_{33}S_{33}^{*\prime -1}S_{23}^{*\prime} - S_{33}'^{-1}S_{23}'S_{22}'^{-1}P_{22}S_{22}^{*\prime}] = X_{12}^{*}.$$

Now let

$$A_{22}' = S_{22}'^{-1}P_{22}S_{22}^{*\prime},\ A_{32}' = A_{33}'S_{33}^{*\prime -1}S_{23}^{*\prime} - S_{33}'^{-1}S_{23}'A_{22}',$$

$$A_{11}' = S_{11}'^{-1}S_{11}^{*\prime},\ A_{12}' = S_{22}'^{-1}(P_{22}S_{12}^{*\prime} - S_{12}'S_{11}'^{-1}S_{11}^{*\prime}),$$

(A.5.10) $$A_{13}' = S_{33}'^{-1}[P_{33}S_{13}^{*\prime} - S_{13}'S_{11}'^{-1}S_{11}^{*\prime} - S_{23}'A_{12}'],$$

$$F_{11} = X_{11}^{*} - (X_{11}A_{11}' + X_{12}A_{12}' + X_{13}A_{13}'),$$

$$F_{22} = X_{22}^{*} - (X_{22}A_{22}' + X_{23}A_{23}'),$$

and

$$F_{21} = X_{21}^{*} - (X_{21}A_{11}' + X_{22}A_{12}' + X_{23}A_{13}').$$

Then it can be easily checked that (Exercise 5.9)

(A.5.11) $$XA + F = X^{*},\ A'VA = V^{*},$$

completing the proof of Proposition 5.1.

To prove Proposition 5.2, note that $\Theta_{12}\Sigma_{22.3}^{-1}\Theta_{12}' = (\Theta_{12}, 0)\begin{pmatrix}\Sigma_{22} & \Sigma_{23}\\ \Sigma_{32} & \Sigma_{33}\end{pmatrix}^{-1}$ $(\Theta_{12}, 0)'$. By setting $X_{13} = 0$ and $X_{23} = 0$ in the above proof and by replacing X_{11}, X_{12}, X_{21} and X_{22} by $\Theta_{11}, \Theta_{12}, \Theta_{21}$ and Θ_{22}, respectively, and V by Σ, the proposition is proved.

Proof of (1) and (2) in Lemma 5.4. (1) Note $U_1 = (X_{12}, X_{13})$, $U_3 = (X_{32}, X_{33})$, $(W_2, W_3) = U_1C' = (X_{12}C_{22}' + X_{13}C_{23}', X_{13}C_{33}')$ and $C(U_1'U_1 + U_3'U_3)C' = I$ where $C = (C_{ij}) \in \mathscr{A}$. Write $U_1'U_1 + U_3'U_3 = C^{-1}C'^{-1} \equiv (H_{ij})$ $(i, j = 2, 3)$. Then $C_{22} = (H_{22} - H_{23}H_{33}^{-1}H_{32})^{-1/2} = H_{22.3}^{-1/2}$ and $C_{23}' = -H_{33}^{-1}H_{32}C_{22}'$, while $H_{22} =$

$X'_{12}X_{12}+V_{22}$, $H_{23}=H'_{32}=X'_{12}X_{13}+V_{23}$ and $H_{33}=X'_{13}X_{13}+V_{33}$. Hence,

$$\begin{aligned} W_2 &= X_{12}C'_{22} + X_{13}C'_{23} = (X_{12} - X_{13}H_{33}^{-1}H_{32})H_{22.3}^{-1/2} \\ &= [X_{12} - X_{13}(X'_{13}X_{13} + V_{33})^{-1}(X'_{13}X_{12} + V_{32})]H_{22.3}^{-1/2} \\ &= [X_{12} - \tilde{X}_{13}(\tilde{X}'_{13}\tilde{X}_{13} + I)^{-1}(\tilde{X}'_{13}X_{12} + V_{33}^{-1/2}V_{32})]H_{22.3}^{-1/2}, \end{aligned} \tag{A.5.12}$$

where $\tilde{X}_{13} = X_{13}V_{33}^{-1/2}$. Here, by using $(\tilde{X}'_{13}\tilde{X}_{13} + I)^{-1} = I - \tilde{X}'_{13}(I + \tilde{X}_{13}\tilde{X}'_{13})^{-1}\tilde{X}_{13}$ and $(I + \tilde{X}_{13}\tilde{X}'_{13})^{-1} = I - (I + \tilde{X}_{13}\tilde{X}'_{13})^{-1}\tilde{X}_{13}\tilde{X}'_{13}$, it follows from (A.5.12) that

(A.5.13)

$$\begin{aligned} W_2 &= [X_{12} - (I + \tilde{X}_{13}\tilde{X}'_{13})^{-1}(\tilde{X}_{13}\tilde{X}'_{13}X_{12} + \tilde{X}_{13}V_{33}^{-1/2}V_{32})]H_{22.3}^{-1/2} \\ &= [(I + \pi_2)^{-1}(X_{12} - X_{13}V_{33}^{-1}V_{32})]H_{22.3}^{-1/2}, \end{aligned}$$

where $\pi_2 = X_{13}V_{33}^{-1}X'_{13}$. On the other hand, using the same relations we get

(A.5.14)

$$\begin{aligned} H_{22.3} &= X'_{12}X_{12} + V_{22} - (X'_{12}X_{13} + V_{23})(X'_{13}X_{13} + V_{33})^{-1}(X'_{13}X_{12} + V_{32}) \\ &= X'_{12}X_{12} + V_{22} - (X'_{12}\tilde{X}_{13} + V_{23}V_{33}^{-1/2}) \\ &\quad \times (\tilde{X}'_{13}\tilde{X}_{13} + I)^{-1}(\tilde{X}'_{13}X_{12} + V_{33}^{-1/2}V_{32}) \\ &= V_{22.3} + Z'Z \qquad \text{with} \qquad Z = (I + \pi_2)^{-1/2}(X_{12} - X_{13}V_{33}^{-1}V_{32}). \end{aligned}$$

Therefore,

$$\operatorname{tr} W'_2W_2 = \operatorname{tr}(I + \pi_2)^{-1}Z(Z'Z + V_{22.3})^{-1}Z',$$

completing the proof of (1).

$$\begin{aligned} \operatorname{tr} W'_3W_3 &= \operatorname{tr} X_{13}(X'_{13}X_{13} + V_{33})^{-1}X'_{13} = \operatorname{tr} \tilde{X}_{13}(\tilde{X}'_{13}\tilde{X}_{13} + I)^{-1}\tilde{X}'_{13} \\ &= \operatorname{tr} \tilde{X}_{13}\tilde{X}'_{13} - \operatorname{tr} \tilde{X}_{13}\tilde{X}'_{13}(I + \tilde{X}_{13}\tilde{X}'_{13})^{-1}\tilde{X}_{13}\tilde{X}'_{13} \\ &= \operatorname{tr} \pi_2(I + \pi_2)^{-1} = n_1 - \operatorname{tr}(I + \pi_2)^{-1}, \end{aligned} \tag{2}$$

which proves (2).

Exercises

5.1. Prove that q satisfies (5.46) whenever either $q(u) = \dfrac{e^{-\operatorname{tr} u/2}}{(2\pi)^{np/2}}$ or $q(u) = \displaystyle\int_0^\infty \frac{e^{-\tau \operatorname{tr} u/2}}{(2\pi)^{np/2}}\tau^{np/2}\,dG(\tau)$.

5.2. Assume $q(u)$ satisfies (5.46). Define $\tilde{q}(v) = \int_{R^{n_2 \times p}} q(v + x'x)\,dx$. Show that $\tilde{q}(v)$ satisfies $\tilde{q}(RvR') = \tilde{q}(v)$ for all $v \in \bar{S}(p)$, $R \in \mathcal{O}(p)$.

5.3. Prove that $\tilde{q}(v)$ is convex if q is also. Hence show that $H(v\,|\,\cdot)$ defined in (5.55) is convex in v.

5.4. Prove that under the model (5.38), the LRT of the hypothesis (5.39) rejects H for large values of $|I_{n_1} + X_1(X'_3X_3)^{-1}X'_1|$.

5.5. Show that $q(u) = e^{-u/2}.\,(2\pi)^{-np/2}$ satisfies Assumption 5.1. Note that in this case $\bar{q}(u)$ is of the same form as $q(u)$.

5.6. Assume $q(u) = \int_0^\infty \frac{e^{-au/2}a^{np/2}}{(\sqrt{2\pi})^{np}}\,dF(a)$ where $u = \operatorname{tr} X'X$, $X\colon n \times p$. Note that $\bar{q}(u)$ is of the same form as $q(u)$. Prove that Assumption 5.1 is satisfied whenever

$$\int_0^\infty a^{i/2}\,dF(a) < \infty, \qquad i = 1, 2, 3.$$

Hints: Clearly (5.76) and (5.78) hold. For (5.77), one needs

$$\iint u^{i/2}e^{-au/2}a^{np/2+i}\,dF(a)\,dX < \infty, \qquad i = 1, 2, 3$$

which is equivalent to the above conditions.

5.7. Refer to Remark 5.3. Show by straightforward calculations that the LRT of the GMANOVA hypothesis (5.58) under the canonical form GMANOVA model (5.56) is given by (5.91).

5.8. Verify (5.96).

5.9. Verify (A.5.11).

Chapter 6 Tests for Covariance Structures

6.1. Introduction

In this chapter we are concerned with the problem of developing optimal invariant tests for certain meaningful structures of the scale matrix Σ of X under the distributional assumption

$$(6.1) \quad X \sim \mathscr{F}_{\mathrm{E}}(M, \Phi \otimes \Sigma), \qquad M \in R^{n \times p}, \qquad \Phi \in \mathscr{S}(n), \qquad \Sigma \in \mathscr{S}(p).$$

Here we recall that $\mathscr{F}_{\mathrm{E}}(M, \Phi \otimes \Sigma)$ denotes the class of elliptically symmetric densities of the form

$$(6.2) \quad f(X \mid M, \Phi, \Sigma) = |\Phi|^{-p/2}|\Sigma|^{-n/2}q(\operatorname{tr}\Sigma^{-1}(X - M)'\Phi^{-1}(X - M)),$$

which was used in Chapter 3. Here q is a function from $[0, \infty)$ into $[0, \infty)$ such that $\int_{R^{n \times p}} q(\operatorname{tr} X'X)\,dX = 1$ (see Chapter 1). The location matrix M must belong to a proper subspace of $R^{n \times p}$ for the (estimability and) testability of the structure of Σ. Throughout the rest of this chapter, we assume that

$$(6.3) \qquad M = \begin{pmatrix} M_1 \\ M_2 \\ 0 \end{pmatrix} \begin{matrix} n_1 \\ n_2, \\ n_3 \end{matrix} \qquad \Phi = I_n, \qquad \text{and} \quad n_3 \geq p.$$

As has been noted in Chapter 3, this specification corresponds to that of a canonical form of the MANOVA problem (see (3.10) of Chapter 3). In other words, we consider the model

$$(6.4)\qquad X = \begin{pmatrix} X_1 \\ X_2 \\ X_3 \end{pmatrix} \begin{matrix} n_1 \\ n_2 \\ n_3 \end{matrix} \sim \mathscr{F}_{\mathrm{E}}\left(\begin{pmatrix} M_1 \\ M_2 \\ 0 \end{pmatrix}, I_n \otimes \Sigma \right) \qquad (n_3 \geq p).$$

Of course, for tests on Σ, writing M as we did in (6.3) rather than as $M = \begin{pmatrix} M^* \\ 0 \end{pmatrix}$ with M^*: $(n_1 + n_2) \times p$ is a matter of formality consistent with the MANOVA problem and does not make any difference in this chapter. Some problems with $\Phi \neq I_n$ associated with outliers are treated in Chapter 7.

As remarked in Section 3.2B of Chapter 3, we claim that any invariant test for any hypothesis on Σ is based on the statistic $X_3'X_3$. This is because the problem remains invariant at least under the group $\mathscr{G} = \mathscr{O}(n_3) \times R^{(n_1+n_2)\times p}$ acting on X by

$$(6.5)\qquad gX = \begin{pmatrix} \begin{pmatrix} X_1 \\ X_2 \end{pmatrix} + F \\ Q_3 X_3 \end{pmatrix} \qquad \text{for} \quad g = (Q_3, F) \in \mathscr{G},$$

and a maximal invariant under this group is $X_3'X_3$. Usually the group $\mathscr{G}$ is enlarged under a specific hypothesis on Σ and, accordingly, a more specific maximal invariant, a suitable function of $X_3'X_3$, is determined. It therefore follows that given a (canonical form of) data matrix X with distribution (6.4), only X_3 is relevant for considering invariant tests on Σ. Hence, below, we restrict our attention to $Z \equiv X_3$ only, which is distributed as

$$(6.6)\qquad Z\colon m \times p \sim \mathscr{F}_{\mathrm{E}}(0, I_m \otimes \Sigma),$$

where $m \equiv n_3$ (see Chapter 2). A density of Z in (6.6) is of the form

$$(6.7)\qquad f(Z \mid \Sigma) = |\Sigma|^{-m/2} q(\operatorname{tr} \Sigma^{-1} Z'Z)$$

for some $q\colon [0, \infty) \to [0, \infty)$ with $\int_{R^{m\times p}} q(\operatorname{tr}(Z'Z))\, dZ = 1$. We are now in a position to describe some specific tests on Σ based on Z. The readers are referred to Section 3.2B of Chapter 3 for a discussion of some commonly used tests for covariance structure under the assumption of normality of Z.

6.2. Testing $\Sigma_{12} = 0$

In this section we consider the problem of testing

$$(6.8)\qquad \mathrm{H}\colon \Sigma_{12} = 0 \qquad \text{versus} \qquad \mathrm{K}\colon \Sigma_{12} \neq 0$$

in the model (6.7), where $Z = (Z_1, Z_2)$, Z_i: $m \times p_i$, $i = 1, 2$, $p_1 + p_2 = p$, and Σ is partitioned as

$$\Sigma = \begin{pmatrix} \Sigma_{11} & \Sigma_{12} \\ \Sigma_{21} & \Sigma_{22} \end{pmatrix}, \quad \Sigma_{ij}: p_i \times p_j \qquad (i, j = 1, 2).$$

It is well known that under the assumption of normality of Z, (6.8) corresponds to the hypothesis of the independence between Z_1 and Z_2, for which several invariant tests are available. However, under $f(Z \mid \Sigma)$ in (6.7) belonging to $\mathscr{F}(0, I_m \otimes \Sigma)$, the hypothesis H is just a special structure of the scale matrix Σ. Of course, if the second moments of Z exist, $\Sigma_{12} = 0$ implies the uncorrelatedness of Z_1 and Z_2. Like the normal case in Section 3.2B of Chapter 3, the problem (6.8) under the distribution (6.7) remains invariant under the group $\mathscr{G} = \mathscr{O}(m) \times \mathrm{GL}(p_1) \times \mathrm{GL}(p_2)$ acting on Z by

$$gZ = (QZ_1C_1', QZ_2C_2') \qquad \text{for} \quad g = (Q, C_1, C_2) \in \mathscr{G}. \tag{6.9}$$

Writing $S = (S_{ij})$ with $S_{ij} = Z_i'Z_j$ $(i, j = 1, 2)$, it follows from (6.7) that S is sufficient for Σ for each $f \in \mathscr{F}(0, I_m \times \Sigma)$ and from (6.9) that a maximal-invariant statistic T and a maximal-invariant parameter Δ, respectively, are the latent roots $d_1 \geq \cdots \geq d_{\min(p_1, p_2)}$ of $S_{12}S_{22}^{-1}S_{21}S_{11}^{-1}$ and the latent roots $\rho_1^2 \geq \cdots \geq \rho_{\min(p_1, p_2)}^2$ of $\Sigma_{12}\Sigma_{22}^{-1}\Sigma_{21}\Sigma_{11}^{-1}$. Without loss of generality, we assume that $p_1 \leq p_2$, and, due to the invariance of the problem, write Σ in the form

$$\Sigma = \begin{pmatrix} I_{p_1} & \Gamma \\ \Gamma' & I_{p_2} \end{pmatrix}, \qquad \Gamma = (\Delta\colon 0)\colon p_1 \times p_2 \tag{6.10}$$

and

$$\Delta = \mathrm{diag}\{\rho_1, \ldots, \rho_{p_1}\}. \tag{6.11}$$

In terms of Δ, the hypotheses in (6.8) are written as

$$\mathrm{H}\colon \Delta = 0 \qquad \text{versus} \qquad \mathrm{K}\colon \Delta \neq 0.$$

Now to apply Wijsman's representation theorem, let $\upsilon = \upsilon_1 \times \upsilon_2 \times \upsilon_3$ be an invariant measure on $\mathscr{G}$, where υ_1 is the invariant probability measure on $\mathscr{O}(m)$ and $d\upsilon_{i+1} = |C_iC_i'|^{-p_i/2}\, dC_i$ $(i = 1, 2)$. Then the probability ratio of T under $\Delta \neq 0$ and $\Delta = 0$ is given by

$$(dP_\Delta^T/dP_0^T)(T(Z)) = H(Z \mid \Delta)/H(Z \mid 0), \tag{6.12}$$

where

$$H(Z \mid \Delta) = \int_{\mathscr{G}} f(gZ \mid \Delta)\chi(g)\upsilon(dg) = \int_{\mathrm{GL}(p_1) \times \mathrm{GL}(p_2)} |\Sigma|^{-m/2} q(\mathrm{tr}\, \Sigma^{-1}CSC') \times |C_1C_1'|^{(m-p_1)/2} |C_2C_2'|^{(m-p_2)/2}\, dC_1\, dC_2, \tag{6.13}$$

where $C = \begin{pmatrix} C_1 & 0 \\ 0 & C_2 \end{pmatrix}$. As has been observed in Chapter 5, an optimum test is obtained by examining the behavior of the ratio as a function of Δ. By using (6.10) with $\Gamma = (\Delta\colon 0)$, we get $|\Sigma| = \Pi_{i=1}^{p_1}(1 - \rho_i^2)$, and (Exercise 6.1)

$$\Sigma^{-1} = \begin{pmatrix} (I - \Gamma\Gamma')^{-1} & -(I - \Gamma\Gamma')^{-1}\Gamma \\ -(I - \Gamma'\Gamma)^{-1}\Gamma' & (I - \Gamma'\Gamma)^{-1} \end{pmatrix} \tag{6.14}$$

$$= \begin{pmatrix} (I - \Delta\Delta')^{-1} & -((I - \Delta\Delta')^{-1}\Delta\colon 0) \\ -\begin{pmatrix} \Delta'(I - \Delta\Delta')^{-1} \\ 0 \end{pmatrix} & \begin{pmatrix} (I - \Delta'\Delta)^{-1} & 0 \\ 0 & I \end{pmatrix} \end{pmatrix},$$

which gives

$$\begin{aligned} \operatorname{tr} \Sigma^{-1}CSC' &= \operatorname{tr}(I - \Delta\Delta')^{-1}C_1S_{11}C_1' \\ &\quad - 2\operatorname{tr}[(I - \Delta\Delta')^{-1}\Delta\colon 0]C_2S_{21}C_1' \\ &\quad + \operatorname{tr}\begin{pmatrix} (I - \Delta'\Delta)^{-1} & 0 \\ 0 & I \end{pmatrix} C_2S_{22}C_2'. \end{aligned} \tag{6.15}$$

Before discussing a general case, we first consider the important particular case of $p_1 = 1$. In this case the testing problem (6.8) reduces to the problem of testing whether the multiple correlation $(\Sigma_{11}^{-1}\Sigma_{12}\Sigma_{22}^{-1}\Sigma_{21})^{1/2}$ between Z_1 and Z_2 is zero. In normal theory, this problem is the well-known R^2-problem, and the UMPI test rejects the null hypothesis whenever the sample multiple correlation coefficient $R = (S_{11}^{-1}S_{12}S_{22}^{-1}S_{21})^{1/2}$ is large (see, e.g., Lehmann, 1959, Chapter 6). Here, by using the argument in Theorem 3.11 of Chapter 3, we shall establish the optimality robustness of the R^2-test under the condition that q in (6.7) is convex. This is similar to the optimality robustness of Hotelling's T^2-test treated in Section 5.3 of Chapter 5.

Theorem 6.1. *When $p_1 = 1$, the R^2-test with critical region $S_{11}^{-1}S_{12}S_{22}^{-1}S_{21} > c$ is UMPI for problem (6.8) whenever q in (6.7) is convex.*

Proof. To evaluate $H(Z \mid \Sigma)$, let $\delta = (\Sigma_{11}^{-1}\Sigma_{12}\Sigma_{22}^{-1}\Sigma_{21})^{1/2}$. Then from (6.15) when $p_1 = 1$, the argument of q in (6.13) is simplified to

$$\begin{aligned} &(1 - \delta^2)^{-1}c_1^2S_{11} - 2\delta(1 - \delta^2)^{-1}C_2^{(1)}S_{21}c_1 \\ &\quad + (1 - \delta^2)^{-1}C_2^{(1)}S_{22}C_2^{(1)\prime} + \sum_{i=2}^{p_2} C_2^{(i)}S_{22}C_2^{(i)\prime} \end{aligned}$$

where $c_1 \equiv C_1$ and $C_2^{(i)}$ is the ith row of C_2 $(i = 1, \ldots, p_2)$. Here, by transforming c_1 into $b_1 = c_1S_{11}^{1/2}$ and $C_2^{(i)}$ into $B_2^{(i)} = C_2^{(i)}S_{22}^{1/2}$, $H(Z \mid \Sigma)$ is evaluated as

$$(6.16)\quad |S_{11}|^{m/2}|S_{22}|^{m/2}\gamma^{m/2}\int_{-\infty}^{\infty}\int_{\mathrm{GL}(p_2)} q\Bigg[\gamma b_1^2 - 2\delta\gamma B_2^{(1)}wb_1 + \gamma B_2^{(1)}B_2^{(1)\prime} + \sum_2^{p_2} B_2^{(i)}B_2^{(i)\prime}\Bigg]|b_1|^{m-1}|B_2B_2'|^{(m-p_2)/2}\,db_1\,dB_2,$$

where $w = S_{22}^{-1/2}S_{21}S_{11}^{-1/2}$, $\gamma = (1-\delta^2)^{-1}$, and $B_2' = (B_2^{(1)\prime},\ldots,B_2^{(p_2)\prime})$. Further, by letting P be a $p_2 \times p_2$ orthogonal matrix with $w/\|w\|$ as its first row and by transforming $B_2^{(i)}$ into $E_2^{(i)} = B_2^{(i)}P$, $H(Z\,|\,\Sigma)$ is evaluated as $k(Z)K(R\,|\,\delta)$ with $k(Z) = |S_{11}|^{m/2}|S_{22}|^{m/2}$ and

$$K(R\,|\,\delta) = \int_{-\infty}^{\infty}\int_{\mathrm{GL}(p_2)} \gamma^{m/2}q\Bigg(\gamma b_1^2 - 2\delta\gamma b_1 e_{11}R + \gamma E_2^{(1)}E_2^{(1)\prime} + \sum_2^{p_2} E_2^{(i)}E_2^{(i)\prime}\Bigg)|b_1|^{m-1}|E_2E_2'|^{(m-p_2)/2}\,db_1\,dE_2,$$

where $E_2' = (E_2^{(1)\prime},\ldots,E_2^{(p_2)\prime})$. In this expression, it is easily shown that $K(-R\,|\,\delta) = K(R\,|\,\delta)$ and $K(R\,|\,\delta)$ is convex because q is convex. Therefore, applying Theorem 3.11 of Chapter 3 with $0 \le R \le 1$ yields the result.

The above result contains the case $p_1 = p_2 = 1$, where R^2 is the square of the sample correlation coefficient $r = S_{12}/(S_{11}S_{22})^{1/2}$ and δ^2 the square of the population correlation coefficient $\rho = \Sigma_{12}/(\Sigma_{11}\Sigma_{22})^{1/2}$.

Corollary 6.1. *Let $p_1 = p_2 = 1$ and assume the convexity of q. Then the test with critical region $|r| > c$ is UMPI for testing $\rho = 0$ versus $\rho \ne 0$.*

For the one-sided problem, the following result holds.

Corollary 6.2. *Let $p_1 = p_2 = 1$ and assume that q is nonincreasing. Then the test with critical region $r > c$ is UMPI for testing $\rho = 0$ versus $\rho > 0$ under the group $\mathcal{G} = R_+ \times R_+$ acting on (Z_1, Z_2) by (c_1Z_1, c_2Z_2) where $(c_1, c_2) \in \mathcal{G}$.*

Proof. By replacing $\mathrm{GL}(p_1) \times \mathrm{GL}(p_2)$ by $R_+ \times R_+$ in $H(Z\,|\,\Delta)$ of (6.13) and by evaluating it in the same way as in the proof of Theorem 6.1, $H(Z\,|\,\Delta)$ is evaluated as

$$H(Z\,|\,\rho) = (S_{11}S_{22})^{m/2}\int_0^{\infty}\int_0^{\infty}\gamma^{m/2}q(\gamma b_1^2 - 2\rho\gamma b_1b_2r + \gamma b_1^2)(b_1b_2)^{m-1}\,db_1\,db_2,$$

(see 6.16), where $\gamma = (1 - \rho^2)^{-1}$. Hence by using the nonincreasingness of q, it is easily observed that the ratio $H(Z\,|\,\rho)/H(Z\,|\,0)$ in (6.12) is a nondecreasing function of r for $\rho > 0$. By the Neyman–Pearson lemma, this implies that the critical region $r > c$ is UMPI, thus completing the proof.

The null robustness of these tests has been observed in Chapter 3. Furthermore, a noninvariant analysis of the problem when $p_1 = p_2 = 1$ has been made in Chapter 4.

We now turn our attention to the general case when $p_1, p_2 > 1$. In this case, as in the case of the MANOVA and GMANOVA problems treated in Chapter 5, what we can expect at best as an optimality property of a test is its being locally best invariant. To derive such a locally best invariant (LBI) test, we make the following assumption.

Assumption 6.1. The function q in (6.7) is thrice continuously differentiable and

$$\begin{aligned}(6.17)\quad &\int |q^{(i)}(\operatorname{tr} C_1C_1' + \operatorname{tr} C_2C_2')|\,|C_1C_1'|^{(m-p_1)/2}|C_2C_2'|^{(m-p_2)/2}\\ &\quad\times |\operatorname{tr} AC_2BC_1'|^{2l+1}[\operatorname{tr}(AA'C_1C_1' + DD'C_2C_2')]^3\,dC_1\,dC_2 < \infty\end{aligned}$$

for any $A\colon p_1 \times p_2, B\colon p_2 \times p_1, D\colon p_2 \times p_1$, $l = 0, 1$ and $i = 1, 2, 3$, where $q^{(i)} = d^iq/dx^i$.

Theorem 6.2. *Under Assumption 6.1, for any invariant test ϕ of size α under $\mathscr{G}$, the power function of ϕ is evaluated as*

$$\begin{aligned}(6.18)\quad \pi(\phi, \Delta) = \alpha + c_1\alpha(\operatorname{tr}\Delta\Delta') + c_2(\operatorname{tr}\Delta\Delta')E_0[\phi(Z)\operatorname{tr} S_{11}^{-1}S_{12}S_{22}^{-1}S_{21}]\\ + o(\operatorname{tr}\Delta\Delta')\end{aligned}$$

uniformly in ϕ, where $\operatorname{tr}\Delta\Delta' = \sum \rho_i^2$ and $c_2 > 0$. Further, the test that rejects H: $\Sigma_{12} = 0$ *when*

$$(6.19)\quad \operatorname{tr} S_{11}^{-1}S_{12}S_{22}^{-1}S_{21} > c$$

is LBI. This test is known as Pillai's test.

Proof. We evaluate the ratio (6.12) for $\|\Delta\|^2 = \sum_1^{p_1} \rho_i^2$ small (see Section 3.5 of Chapter 3). Note that

$$\begin{aligned}(6.20)\quad (I - \Delta\Delta')^{-1} = I + \Delta\Delta' + o(\|\Delta\|^2) \quad\text{and}\quad (I - \Delta\Delta')^{-1}\Delta\\ = \Delta + o(\|\Delta\|^2).\end{aligned}$$

Then $\psi(C\colon \Delta) = \operatorname{tr} \Sigma^{-1} CSC'$, which is the argument of q in (6.13), is expanded as

$$\delta(C\colon \Delta) \equiv \psi(C\colon \Delta) - \psi(C\colon 0) = h(C\colon \Delta) + o(\|\Delta\|^2)h(C\colon \Delta),$$

where $C = \begin{pmatrix} C_{12} \\ C_{22} \end{pmatrix}$, $W = S_{22}^{-1/2} S_{21} S_{11}^{-1/2}$, and

$$h(C\colon \Delta) = \operatorname{tr} \Delta\Delta' C_1 C_1' - 2 \operatorname{tr} \Delta C_{12} W C_1' + \operatorname{tr} C_{12} C_{12}'.$$

Therefore, by Assumption 6.1, $q(\operatorname{tr} \Sigma^{-1} CSC') = q(\psi(C\colon \Delta))$ is expanded as

$$q(\operatorname{tr} C_1 C_1' + \operatorname{tr} C_2 C_2') + \delta(C\colon \Delta) q^{(1)}(\operatorname{tr} C_1 C_1' + \operatorname{tr} C_2 C_2')$$
$$+ \tfrac{1}{2}\delta(C\colon \Delta)^2 q^{(2)}(\operatorname{tr} C_1 C_1' + \operatorname{tr} C_2 C_2')$$
$$+ \tfrac{1}{6}\delta(C\colon \Delta)^3 q^{(3)}(\operatorname{tr} C_1 C_1' + \operatorname{tr} C_2 C_2' + (1-\alpha)\delta(C\colon \Delta)) \qquad (0 < \alpha < 1).$$

To evaluate the integrals of these terms over $\mathrm{GL}(p_1) \times \mathrm{GL}(p_2)$, note that the integrals of odd functions of either C_1 or C_2 are zero because the integrals are finite by Assumption 6.1. In other words, with $\eta(dC) = \Pi_{i=1}^{2} |C_i C_i'|^{(m-p_i)/2}\, dC_i$,

$$\int [\operatorname{tr} \Delta\Delta'(C_1 C_1' + C_{12} C_{12}')]^k (\operatorname{tr} \Delta C_{12} W C_1')^{2j+1} \times q^{(i)}(\operatorname{tr} C_1 C_1' + \operatorname{tr} C_2 C_2')\eta(dC) = 0, \tag{6.21}$$

where $i = 1, 2, 3$, $j = 0, 1, 2$, $k = 1, 2$. It is also noted that the elements of W are all less than unity whatever the data matrix is. On the basis of these observations and by proceeding as in Section 5.4 of Chapter 5, we conclude that the ratio (6.12) is evaluated as

$$1 + \operatorname{tr}(\Delta\Delta')K_1 + \frac{2}{D_0}\int (\operatorname{tr} \Delta C_{12} W_2 C_1')^2 q^{(2)}(\operatorname{tr} CC')\eta(dC) + o(\|\Delta\|^2), \tag{6.22}$$

where $\operatorname{tr} CC' = \operatorname{tr} C_1 C_1' + \operatorname{tr} C_2 C_2'$, K_1 is a constant, $D_0 = \int q(\operatorname{tr} CC')\eta(dC)$, and $o(\|\Delta\|^2)$ is uniform in Z. For example, it is easy to show (Exercise 6.2):

$$\int [\operatorname{tr} \Delta\Delta'(C_1 C_1' + C_{12} C_{12}')] q^{(1)}(\operatorname{tr} CC')\eta(dC) = c_1 \operatorname{tr} \Delta\Delta' \qquad \text{for some} \quad c_1, \tag{6.23}$$

$$\int [\operatorname{tr} \Delta\Delta'(C_1 C_1' + C_{12} C_{12}')]^2 q^{(2)}(\operatorname{tr} CC')\eta(dC) = o(\|\Delta\|^2), \tag{6.24}$$

$$\int [\operatorname{tr} \Delta\Delta'(C_1 C_1' + C_{12} C_{12}')][\operatorname{tr} \Delta C_{12} W C_1']^2 q^{(3)}(\operatorname{tr} CC')\eta(dC) = o(\|\Delta\|^2) \tag{6.25}$$

uniformly in Z, and so on. To evaluate (6.22) further, we need

$$(6.26)\quad \int (\operatorname{tr} \Delta C_{12} W C_1')^2 q^{(2)}(\operatorname{tr} CC')\eta(dC) = K_2(\operatorname{tr} \Delta\Delta')(\operatorname{tr} S_{22}^{-1}S_{21}S_{11}^{-1}S_{12})$$

$$\text{for some}\quad K_2\ (>0).$$

We leave this as an exercise (Exercise 6.5; see the case of the GMANOVA problem). Therefore, noting that the null distribution of a maximal invariant does not depend on q as well as on Δ, the power function of an invariant test φ is given by

$$(6.27)\quad \pi(\varphi, \Delta) = \alpha(1 + K_1 \operatorname{tr} \Delta\Delta') + (2K_2/D_0)E_0[\varphi \operatorname{tr} S_{22}^{-1}S_{21}S_{11}^{-1}S_{12}] + o(\|\Delta\|^2).$$

Since $K_2/D_0 > 0$, maximizing the second term of the right side yields the desired result.

It may be mentioned that Schwartz (1967a) proved the LBI property of the Pillai test under the assumption of normality.

6.3. Testing Sphericity

Another common problem of covariance structure in applications is testing the sphericity of Σ in the model (6.7):

$$(6.28)\quad \mathrm{H}: \Sigma = \sigma^2 I_p \quad \text{versus} \quad \mathrm{K}: \Sigma = \sigma^2\Omega, \quad \Omega \neq I.$$

Continuing from Application 3.9 in Chapter 3, in this section we establish the optimality robustness of the LBI test derived by Sugiura (1972) under the normality assumption on $Z: m \times p$. Recall that the test rejects H for large values of $\operatorname{tr} S^2/(\operatorname{tr} S)^2$. As has been discussed in Chapter 3, the problem remains invariant under the group $\mathscr{G} = R_+ \times \mathscr{O}(p)$ acting on Z by $g(Z) = cZQ'$ for $g = (c, Q) \in \mathscr{G}$. Then by Theorem 3.7 in Chapter 3, the probability ratio R of a maximal invariant $T = T(Z)$, evaluated at $T = T(Z)$, is given by

$$(6.29)\quad R = |\Omega|^{-m/2} \int_{\mathscr{O}(p)} [1 + F]^{-mp/2} \nu_2(dQ),$$

where $F = \operatorname{tr}(\Omega^{-1} - I)QSQ'/\operatorname{tr} S$ with $S = Z'Z$ and ν_2 is the invariant probability measure on $\mathscr{O}(p)$. Recall that the null distribution of T is independent of q and σ^2. Replacing Q by ΓQ with $\Gamma \in \mathscr{O}(p)$ leaves the integral the same, and hence Ω is assumed to be of the form

$$\Omega = \operatorname{diag}\{w_1, \ldots, w_p\}. \tag{6.30}$$

The null hypothesis in this form reduces to H: $w_1 = \cdots = w_p = 1$, so that local alternatives correspond to $|w_i^{-1} - 1|$ being small for all i but not zero for all i. To evaluate R locally, expand the integrand as

$$1 - aF + a(a+1)F^2/2 - a(a+1)(a+2)F^3(1+\theta F)^{-a-3}/6, \tag{6.31}$$

where $0 < \theta < 1$ and $a = mp/2$. The integrals of the F^i's over $\mathcal{O}(p)$ are evaluated as

$$\int_{\mathcal{O}(p)} F v_2(dQ) = \operatorname{tr}(\Omega^{-1} - I)/p \tag{6.32}$$

from $\int_{\mathcal{O}(p)} \operatorname{tr} AQBQ' v_2(dQ) = \operatorname{tr} A \operatorname{tr} B/p$ (see Lemma 2.5 in Chapter 2);

$$\int_{\mathcal{O}(p)} F^2 v_2(dQ) = 3[\operatorname{tr}(\Omega^{-1} - I)^2][\operatorname{tr} S^2]/[p(p+1)(\operatorname{tr} S)^2] \tag{6.33}$$

from $\int_{\mathcal{O}(p)} (\operatorname{tr} AQBQ')^2 v_2(dQ) = 3 \operatorname{tr} A^2 \operatorname{tr} B^2/p(p+1)$ (see Section 2.1 of Chapter 2 and Exercise 6.6);

$$\int_{\mathcal{O}(p)} F^3(1+\theta F)^{-a-3} v_3(dQ) = o(\operatorname{tr}(\Omega^{-1} - I)^2) \qquad \text{uniformly in } Z, \tag{6.34}$$

because $|F|$ is bounded above by $\sum |w_i^{-1} - 1|$ from $\Omega^{-1} - I \leq (\sum |w_i^{-1} - 1|)I$. Therefore, the ratio R is evaluated as

(6.35)
$$R = 1 - a \operatorname{tr}(\Omega^{-1} - I) + 3a(a+1)[\operatorname{tr} S^2/(\operatorname{tr} S)^2][\operatorname{tr}(\Omega^{-1} - I)^2]/2p(p+1) + o(\operatorname{tr}(\Omega^{-1} - I)^2)$$

uniformly in Z. Thus we obtain

Theorem 6.3. *For problem* (6.28) *in model* (6.7), *the power function of an invariant test* ϕ *of size* α *is evaluated as*

$$\pi(\phi, (\sigma^2, \Omega)) = \alpha\left[1 - \frac{m}{2}\gamma\right] + K\gamma^2 E_0[\phi \operatorname{tr} S^2/(\operatorname{tr} S)^2] + o(\gamma^2),$$

where $K = 3m(m+2)/8p(p+1)$. *Furthermore, the test with critical region* $U \equiv \operatorname{tr} S^2/(\operatorname{tr} S)^2 > c$ *is LBI whatever* q *in* (6.7) *may be, and the null and non-null distributions of* U *are the same as those under normality.*

The latter part follows from a result in Section 3.4 of Chapter 3. In fact, any invariant test is null and nonnull robust.

6.4. Testing $\Sigma = I$ versus $\Sigma > I$

In this section the problem of testing H: $\Sigma = I$ versus K: $\Sigma > I$ is considered under model (6.7). As mentioned in Section 3.2(B3) of Chapter 3, when Z is distributed as $N(0, I_n \otimes \Sigma)$, John (1971) derived an LBI test with critical region $\operatorname{tr} S > c$, where $S = Z'Z$: $p \times p$ and $n \geq p$. We shall show that it is also LBI under the pdf (6.7), provided q satisfies certain conditions specified below. First note that the problem remains invariant under the group $\mathscr{G} \equiv \mathscr{O}(p)$ acting on the left of Z by $Z \to Zg'$ for $g \in \mathscr{G}$. From (3.47) and (3.48) of Chapter 3, the ratio R is evaluated as

$$(6.36) \qquad R = dP_\Sigma^T/dP_I^T = \int_{\mathscr{G}} |\Sigma|^{-n/2} q(\operatorname{tr}\Sigma^{-1} gSg')\nu(dg)/q(\operatorname{tr} S),$$

where P_Σ^T is the distribution of a maximal invariant under Σ, and ν is the invariant probability measure on $\mathscr{G}$. However, P_I^T here does depend on q, which implies that the null distribution of T depends on q. Choosing $T = \operatorname{diag}\{t_1, \ldots, t_p\}$, where $t_1 > \cdots > t_p$ are the latent roots of S, it is well known (e.g., Anderson 1958, p. 318) that

$$(6.37) \qquad dP_I^T = aq(\operatorname{tr} T)J(T)\,dT,$$

where $J(T) = [\Pi_{i>j}(t_i - t_j)]$, $dT = \Pi\, dt_i$, and a is a constant. Hence from (6.36) and (6.37), the pdf of T under Σ with respect to dT is given by

$$(6.38) \qquad h(T\,|\,\Sigma) = aJ(T)\int_{\mathscr{O}(p)} |\Sigma|^{-n/2} q(\operatorname{tr}\Sigma^{-1} gTg')\nu(dg),$$

from which $h(T\,|\,\Sigma)/h(T\,|\,I) = R$ is observed. To evaluate (6.38) for Σ close to I, we make

Assumption 6.2. $q \in \mathscr{Q}$, where $\mathscr{Q}$ is the class of twice continuously differentiable functions from $[0, \infty)$ into $[0, \infty)$ such that $\int_{R^{np}} q(Z'Z)\,dZ = 1$,

$$(6.39) \qquad \int (\operatorname{tr} T)^i |q^{(i)}(\operatorname{tr} T)| J(T)\,dT < \infty \qquad (i = 1, 2),$$

$$(6.40) \qquad q^{(2)}(x) \geq 0 \quad \text{for all} \quad x, \quad \text{and} \quad q^{(2)} \text{ is nonincreasing},$$

where $q^{(i)}(x) = d^i q(x)/dx^i$.

Then proceeding as in the problems in Sections 6.2 and 6.3, we obtain

Theorem 6.4. *When Assumption* 6.2 *holds and when* $xq^{(1)}(x)/q(x)$ *is decreasing in* x, *the test with critical region* $\operatorname{tr} S > c$ *is LBI under the* pdf (6.7). *But the null distribution of* $\operatorname{tr} S$ *depends on* $q \in \mathcal{Q}$.

The proof is similar to that of Theorem 6.3, and here we simply sketch it. First expand $|\Sigma^{-1}|^{n/2}$ as $1 - \frac{n}{2}\delta + o(\delta)$ with $\delta = \operatorname{tr}(\Sigma^{-1} - I)$ and $q(\operatorname{tr}\Sigma^{-1}gTg')$ in (6.38) as

(6.41) $$q(\operatorname{tr} T) + [\operatorname{tr}(\Sigma^{-1} - I)gTg']q^{(1)}(\operatorname{tr} T) + \tfrac{1}{2}[\operatorname{tr}(\Sigma^{-1} - I)gTg']^2 q^{(2)}(z),$$

where $z = \operatorname{tr} T + c\operatorname{tr}(\Sigma^{-1} - I)gTg'$ for some $0 \le c \le 1$. Then from $z \ge \operatorname{tr} T(1 + \delta)$, the integration of the third term in (6.41) is evaluated from (6.40) as

(6.42)
$$\left|\frac{1}{2}\int [\operatorname{tr}(\Sigma^{-1} - I)gTg']^2 q^{(2)}(z)\nu(dg)\right| \le \tfrac{1}{2}\delta^2 \operatorname{tr} T^2 q^{(2)}(\operatorname{tr} T[1 + \delta]) \equiv \delta^2 H(T, \delta).$$

Here from (6.39), $\sup_{-1/2 < \delta \le 0} \int [H(T, \delta)/q(\operatorname{tr} T)]h(T \mid I)\,dT < \infty$. Hence, since $\int_{\mathcal{O}(p)} \operatorname{tr} AgBg'\nu(dg) = \operatorname{tr} A \operatorname{tr} B/p$ (see Lemma 2.5 with $k = 1$), the ratio $h(T \mid \Sigma)/h(T \mid I)$ is

(6.43) $$R = 1 + \delta\left[\{\operatorname{tr} Tq^{(1)}(\operatorname{tr} T)/pq(\operatorname{tr} T)\} - \frac{n}{2}\right] + o_1(\delta, T).$$

Under Assumption 6.2, $\int [o_1(\delta, T)]h(T \mid I)\,dT = o(\delta)$. This implies for any invariant test $\phi(T)$ of level α, that the power function is evaluated as (see 3.95 of Chapter 3)

(6.44) $$\pi(\phi, \Sigma) = \alpha + \delta E_I\left\{\phi(T)\left[\{\operatorname{tr} Tq^{(1)}(\operatorname{tr} T)/pq(\operatorname{tr} T)\} - \frac{n}{2}\right]\right\} + o(\delta).$$

Therefore, since $\Sigma > I$ implies $\delta < 0$, and since $xq^{(1)}(x)/q(x)$ is decreasing in x, the test based on $\operatorname{tr} T > c$ maximizes the second term of (6.44), completing the proof.

Lemma 6.1. *If* q *is twice differentiable and logconcave,* $xq^{(1)}(x)/q(x)$ *is decreasing.*

The proof is left as an exercise (Exercise 6.7).

Exercises

6.1. Prove (6.14). [Hint: The first equality is standard and can be verified, for example, by direct multiplication. The second equality follows upon writing $\Gamma = (\Delta\colon 0)$.]

6.2. Prove (6.23).

6.3. Use (6.23) and the fact that $\operatorname{tr} \Delta\Delta'(C_1 C_1' + C_{12} C_{12}') \leq (\operatorname{tr} \Delta\Delta')(\operatorname{tr}(C_1 C_1' + C_{12} C_{12}'))$ to prove (6.24).

6.4. Use (6.23) and the fact that $\operatorname{tr}[\Delta C_{12} W C_1']^2 \leq (\operatorname{tr} \Delta\Delta')(\operatorname{tr} C_{12}' C_{12} W C_1' C_1)$ and the elements of W are bounded by unity in absolute value to prove (6.25).

6.5. Proceeding as in the GMANOVA problem, prove (6.26).

6.6. Prove (6.33).

6.7. Prove Lemma 6.1.

Chapter 7 Detection of Outliers

7.1. Introduction

The identification of outlying or spurious or aberrant observations in a set of data, observations that are far removed from the main group, is a very old and common problem in statistics. It is also a difficult problem because, quite generally, a few genuine values may be far off from the rest and therefore may seem erroneous. Although this phenomenon may not occur frequently, there is a positive probability that such values will occur in an experiment. The question then arises if indeed we are always sampling from the same population or the population is undergoing some changes. Associated with the problem of detection and identification of possible outliers in the data is the equally difficult and dual problem of statistical inference. This involves estimating a set of parameters or testing a hypothesis, assuming the presence of some outliers in the data. It is thus vital that outliers in the data, if any, be detected with powerful techniques because the presence or absence of such aberrant observations indeed makes a major difference in the subsequent analysis of the data.

As mentioned in Chapter 3, our goal here is to develop appropriate statistical tests for the detection of outliers. We are not concerned with the tasks of determining exactly which observations, if any, found by the tests are outliers and what to do with them. The reader is referred to the books by

Barnett and Lewis (1978), Hawkins (1980). Gnanadesikan (1977), and Sinha (1986) for a discussion of these latter aspects.

Mathematical models providing a structure to the outlier problem were formally introduced in Section 3.2 of Chapter 3. Typically, as mentioned there, the change in the spurious observation is hypothesized as a shift in either the location or the scale through an unknown scalar parameter (see equations 3.21 and 3.22 in Section 3.2 of Chapter 3). The parameter, if nonzero, signifies the presence of some outliers in the data.

It was Ferguson (1961) who first undertook a systematic study of the detection of outliers in the univariate normal case based on mean and variance slippage models. The approach adopted by Ferguson is traditional and rests on finding an explicit expression of a maximal invariant and then its null and nonnull distributions. Curiously, as Ferguson found out, the sample kurtosis b_2 is the locally best invariant test statistic for both the mean and variance slippage.

The extensions to the multivariate case dealing with the problem of testing H: $\Delta = 0$ versus K: $\Delta \neq 0$ under (3.21) of Chapter 3 for mean slippage and the problem of testing H: $\Delta = 0$ versus K: $\Delta > 0$ under (3.22) of Chapter 3 for dispersion slippage, which is the subject matter of this chapter, have been seriously attempted. Schwager and Margolin (1982) have shown that the test based on Mardia's (1970) multivariate kurtosis statistic $b_{2,p}$ is locally best invariant for the multivariate mean slippage problem under normality. The optimality robustness of the result was shown by Sinha (1984). On the other hand, for the multivariate dispersion slippage problem, Das and Sinha (1986) showed that the test based on the same kurtosis statistic $b_{2,p}$ is locally best invariant in the class of spherical distributions and that the test is nonnull-robust.

Section 2 of this chapter deals with the mean slippage case and Section 3 with the case of dispersion slippage. Of course, Ferguson's univariate normal result is a special case.

7.2. Test for Mean Slippage

In this section, we are concerned with the problem of developing optimum invariant tests for the detection of outliers with mean slippage. The framework is based on a data matrix $X: n \times p$ under the distributional assumption

$$(7.1) \qquad X \sim \mathscr{F}_E(\mu, I_n \otimes \Sigma), \qquad \mu \in R^{n \times p}, \qquad \Sigma \in \mathscr{S}(p),$$

where $\mathscr{F}_E(\mu, I_n \otimes \Sigma)$ denotes the class of elliptically symmetric densities as in (6.1) of Chapter 6. The location matrix μ is assumed to follow the structure

$$\mu = 1\alpha' + \Delta A \Sigma^{1/2} \tag{7.2}$$

as given in (3.21) of Chapter 3. Under the null hypothesis H: $\Delta = 0$, the different rows of X are homogeneous without presence of any outlier. However, under the alternative K: $\Delta \neq 0$, rows of X corresponding to non-null rows of A represent the outlying observations. Here A is a specified matrix except for the permutation of its rows, so that rejection of H does not identify the outlying observations. Throughout this chapter, we assume $n \geq p + 1$.

The above model (7.1) resembles the model (3.5) in Section 3.2 of Chapter 3, except that Φ is taken as I_n and the normality there is replaced by the class of elliptically symmetric densities of the form

$$f(X \mid \mu, \Sigma) = |\Sigma|^{-n/2} q(\operatorname{tr} \Sigma^{-1}(X - \mu)'(X - \mu)), \tag{7.3}$$

where q is a function from $[0, \infty)$ into $[0, \infty)$ such that $\int_{R^{n \times p}} q(\operatorname{tr} X'X)\, dX = 1$ (see Chapter 1).

To derive an optimum invariant test, we note that the problem remains invariant under the group

$$\mathscr{G} = \mathscr{G}_1 \times \mathscr{G}_2 \times \mathscr{G}_3 = \mathrm{GL}(p) \times \mathscr{P} \times R^p$$

acting on the left of X as $X \to g(X) = \Gamma X C + 1\eta'$ for $g = (C, \Gamma, \eta) \in \mathscr{G}$. Here, as mentioned in Application 3.11 in Chapter 3, $\mathscr{P}$ represents the finite group of permutations of the n rows of A. The group operation of two elements $g_1 = (C_1, \Gamma_1, \eta_1)$, $g_2 = (C_2, \Gamma_2, \eta_2)$ in $\mathscr{G}$ is defined by

$$g_1 \cdot g_2 = (C_1 C_2, \Gamma_1 \Gamma_2, C_1 \eta_2 + \eta_1).$$

Then, as in (3.47) or (3.83) of Chapter 3, the probability ratio R of a maximal invariant $T = T(X)$ is given by

$$R = R(\pi(X)) \equiv (dP^T_{\theta_1}/dP^T_{\theta_0})(T(X)) = H(X \mid \theta_1)/H(X \mid \theta_0) \tag{7.4}$$

with

$$H(X \mid \theta) = \int_{\mathscr{G}} f(g(X) \mid \theta) \chi(g) \nu(dg), \tag{7.5}$$

where χ is a left multiplier function, and ν is an invariant measure on $\mathscr{G}$ and $\theta_i \in H_i$, $i = 0, 1$; $\theta = (\alpha, \Delta, \Sigma)$. Note that $P^T_{\theta_0}$ is independent of unknown parameters. As a left-invariant measure ν on $\mathscr{G}$, we take $\nu = \nu_1 \times \nu_2$ where ν_1 is the discrete uniform probability measure on $\mathscr{P}$ with mass $1/n!$ at each of the points $\Gamma_\alpha \in \mathscr{P}$, and $d\nu_2(C, \eta) = d\eta\, dC/|C'C|^{(p+1)/2}$ with $d\eta$, dC = Lebesgue on R^p and R^{p^2}, respectively. Moreover, by the invariance of the problem, we can take $\alpha = 0$ and $\Sigma = I_p$ while computing the ratio R so that, for specific A, Δ is a maximal invariant parameter. Finally, with $\chi(g)$ being the inverse of the Jacobian of transformation $X \to g(X)$, we get $\chi(g) = |C'C|^{n/2}$. By denoting the ratio R by $R_\Delta(T(X))$, we then have the following result.

Lemma 7.1.

$$R_\Delta(T(X)) = \sum_\alpha \int_{\mathscr{G}_1} \tilde{q}[\operatorname{tr}(C'C - 2\Delta C'\tilde{\tilde{x}}'_\alpha A + \Delta^2 \tilde{\tilde{A}})]|C'C|^{(n-p-1)/2}\, dC$$

$$\div \sum_\alpha \int_{\mathscr{G}_1} \tilde{q}[\operatorname{tr} C'C]|C'C|^{(n-p-1)/2}\, dC,$$

where $\tilde{q}(u) = \int_{R^p} q(u + nv'v)\, dv$, $u \geq 0$, $\tilde{\tilde{x}}'_\alpha = S^{-1/2}(\Gamma_\alpha X - \underset{\sim}{1}\bar{X}')'$, $\bar{X} = X'\underset{\sim}{1}/n$ *and* $\tilde{\tilde{A}} = A'(I_n - \underset{\sim}{1}\underset{\sim}{1}'/n)A$, *with* A *corresponding to a fixed permutation* $(1, \ldots, n)$ *of its rows. Here* $S = \sum_1^n (X_i - \bar{X})(X_i - \bar{X})'$, $S^{-1/2} = (S^{1/2})^{-1}$, *and* $S^{1/2}$ *is the positive square root of* S. *Also*, $X = (X_1, \ldots, X_n)'$.

Proof. By taking $\mu = \Delta A$ and $\Sigma = I_p$ in (7.3), and by replacing X by $g(X) = \Gamma X C + 1\eta'$, it follows that

$$f(g(X)\,|\,\Delta) = q\{n\eta^{*\prime}\eta^* + \operatorname{tr}[C'SC - 2\Delta C'(\Gamma X - \underset{\sim}{1}\bar{X}')'A + \Delta^2 \tilde{\tilde{A}}]\}, \tag{7.6}$$

where $\eta^{*\prime} = \eta' - \Delta \underset{\sim}{1}'A/n + \bar{X}'C$, and we have used the fact that $X'\Gamma'\Gamma X = X'X$ for all $\Gamma \in \mathscr{P}$. Integrating out η^* in q with respect to the Lebesgue measure, summing over Γ in $\mathscr{P}$, and making the transformation from C to $S^{1/2}C$, yields

(7.7)

$$H(X\,|\,\Delta) = \frac{|S|^{-(n-1)/2}}{n!} \sum_\alpha \int \tilde{q}[\operatorname{tr}\{C'C - 2\Delta C'\tilde{\tilde{x}}'_\alpha A + \Delta^2 \tilde{\tilde{A}}\}]|C'C|^{(n-p-1)/2}\, dC.$$

Since the null hypothesis corresponds to $\Delta = 0$, taking the ratio $H(X\,|\,\Delta)/H(X\,|\,0)$ completes the proof.

We now proceed as in Section 3.5 of Chapter 3. There it is shown that, under Assumption 3.3, the optimality robustness of the UMPI property of the test based on $v(X)$ holds (Theorem 3.11). In our problem here, the function

$H(X \mid \Delta)$ does not satisfy Assumption 3.3 of Chapter 3. This is similar to the Example 3.6 illustrated in Section 3.5 of Chapter 3.

To establish the robustness of an LBI property, we follow the general procedure outlined in Section 3.5 of Chapter 3. This involves an expansion of $R_\Delta(T(X))$ around $\Delta = 0$, as in (3.94) of Section 3.5 of Chapter 3, through an initial expansion of the integrand of $H(X \mid \Delta)$ in (7.7), similar to (3.93) of Section 3.5 in that chapter. The necessary regularity conditions for this purpose are stated below.

Assumption 7.1. Assume that $\tilde{q}(\cdot)$ satisfies the following:

(i) $\tilde{q}(x + y)$ admits the continuous fourth derivative in y for every fixed x,
(ii) $\int_{\mathscr{G}_1} c_{11}^4 |\tilde{q}^{(i)}(\operatorname{tr} C'C)||C'C|^{(n-p-1)/2}\, dC < \infty$,
$\int_{\mathscr{G}_1} c_{12}^4 |\tilde{q}^{(i)}(\operatorname{tr} C'C)||C'C|^{(n-p-1)/2}\, dC < \infty, \qquad i = 1, 2, 3, 4,$

where $\tilde{q}^{(i)}(u) = d^i\tilde{q}(u)/du^i$.

Denote the numerator of $R_\Delta(T(X))$ in Lemma 7.1 by $H^*(X \mid \Delta)$. A Taylor expansion of $H^*(X \mid \Delta)$ around $\Delta = 0$, under Assumption 7.1, is then given by

(7.8)

$$
\begin{aligned}
H^*(X \mid \Delta) = H^*(X \mid 0) + \sum_\alpha \int_{\mathscr{G}_1} \Big[& (-2\Delta \operatorname{tr} C'\tilde{\tilde{x}}'_\alpha A + \Delta^2 \operatorname{tr} \tilde{\tilde{A}})\tilde{q}^{(1)}(\operatorname{tr} C'C) \\
& + \tfrac{1}{2}(-2\Delta \operatorname{tr} C'\tilde{\tilde{x}}'_\alpha A + \Delta^2 \operatorname{tr} \tilde{\tilde{A}})^2\tilde{q}^{(2)}(\operatorname{tr} C'C) \\
& + \tfrac{1}{6}(-2\Delta \operatorname{tr} C'\tilde{\tilde{x}}'_\alpha A + \Delta^2 \operatorname{tr} \tilde{\tilde{A}})^3\tilde{q}^{(3)}(\operatorname{tr} C'C) \\
& + \tfrac{1}{24}(-2\Delta \operatorname{tr} C'\tilde{\tilde{x}}'_\alpha A + \Delta^2 \operatorname{tr} \tilde{\tilde{A}})^4\tilde{q}^{(4)}(\operatorname{tr} C'C) \\
& + \tfrac{1}{24}(-2\Delta \operatorname{tr} C'\tilde{\tilde{x}}'_\alpha A + \Delta^2 \operatorname{tr} \tilde{\tilde{A}})^4 \\
& \times \{\tilde{q}^{(4)}(\operatorname{tr} C'C + \varepsilon(-2\Delta \operatorname{tr} C'\tilde{\tilde{x}}'_\alpha A + \Delta^2 \operatorname{tr} \tilde{\tilde{A}})) \\
& - \tilde{q}^{(4)}(\operatorname{tr} C'C)\} \Big] |C'C|^{(n-p-1)/2}\, dC,
\end{aligned}
$$

for some $0 < \varepsilon < 1$.

The quantity $H^*(X \mid 0)$ represents the denominator of $R_\Delta(T(X))$. The different terms in the right-hand side (rhs) of (7.8) are now evaluated. Many terms vanish due to the fact that

(7.9)

$$
\int_{\mathscr{G}_1} (\operatorname{tr} C'\tilde{\tilde{x}}'_\alpha A)^{2j+1}\tilde{q}^{(i)}(\operatorname{tr} C'C)|C'C|^{(n-p-1)/2}\, dC = 0, \qquad j = 0, 1, \quad i = 1, 2, 3, 4,
$$

which follows from Assumption 7.1 upon changing C to $-C$. Clearly the coefficients of odd powers of Δ vanish when all the terms except the last in the rhs of (7.8) are considered, and the coefficients ξ_{2i} of Δ^{2i}, $i = 1, 2, 3, 4$, collected from all these terms are given by

$$\text{(7.10)} \quad \begin{aligned} \xi_2 &= (\operatorname{tr} \tilde{\tilde{A}}) \sum_\alpha \int_{\mathscr{G}_1} \tilde{q}^{(1)}(\operatorname{tr} C'C)|C'C|^{(n-p-1)/2}\, dC \\ &\quad + 2 \sum_\alpha \int_{\mathscr{G}_1} (\operatorname{tr} C'\tilde{\tilde{x}}'_\alpha A)^2 \tilde{q}^{(2)}(\operatorname{tr} C'C)|C'C|^{(n-p-1)/2}\, dC, \end{aligned}$$

$$\text{(7.11)} \quad \begin{aligned} \xi_4 &= \tfrac{1}{2}(\operatorname{tr} \tilde{\tilde{A}})^2 \sum_\alpha \int_{\mathscr{G}_1} \tilde{q}^{(2)}(\operatorname{tr} C'C)|C'C|^{(n-p-1)/2}\, dC \\ &\quad + 2(\operatorname{tr} \tilde{\tilde{A}}) \sum_\alpha \int_{\mathscr{G}_1} (\operatorname{tr} C'\tilde{\tilde{x}}'_\alpha A)^2 q^{(3)}(\operatorname{tr} C'C)|C'C|^{(n-p-1)/2}\, dC \\ &\quad + \tfrac{2}{3} \sum_\alpha \int_{\mathscr{G}_1} (\operatorname{tr} C'\tilde{\tilde{x}}'_\alpha A)^4 \tilde{q}^{(4)}(\operatorname{tr} C'C)|C'C|^{(n-p-1)/2}\, dC, \end{aligned}$$

$$\text{(7.12)} \quad \begin{aligned} \xi_6 &= \tfrac{1}{6}(\operatorname{tr} \tilde{\tilde{A}})^3 \sum_\alpha \int_{\mathscr{G}_1} \tilde{q}^{(3)}(\operatorname{tr} C'C)|C'C|^{(n-p-1)/2}\, dC \\ &\quad + (\operatorname{tr} \tilde{\tilde{A}})^2 \sum_\alpha \int_{\mathscr{G}_1} (\operatorname{tr} C'\tilde{\tilde{x}}'_\alpha A)^2 q^{(4)}(\operatorname{tr} C'C)|C'C|^{(n-p-1)/2}\, dC, \end{aligned}$$

$$\text{(7.13)} \quad \xi_8 = \tfrac{1}{24}(\operatorname{tr} \tilde{\tilde{A}})^4 \sum_\alpha \int_{\mathscr{G}_1} \tilde{q}^{(4)}(\operatorname{tr} C'C)|C'C|^{(n-p-1)/2}\, dC.$$

Evidently ξ_8 is a constant. The following result, whose proof is given in Appendix 7, shows that ξ_2, ξ_6, and the first two terms of ξ_4 are all constants.

Lemma 7.2.

$$\sum_\alpha (\operatorname{tr} C'\tilde{\tilde{x}}'_\alpha A)^2 = n(n-2)!(\operatorname{tr} AC'CA') - (n-2)!(1'AC'CA'1).$$

To evaluate the last term of ξ_4, we need the following result whose proof is also deferred to Appendix 7.

Lemma 7.3.

$$\begin{aligned} \sum_\alpha (\operatorname{tr} C'\tilde{\tilde{x}}'_\alpha A)^4 = (n-4)![&(n^3+n^2)T_1 + (3n^2-9n+3)T_2 \\ &- (3n^2-3n-6)T_3 - (3n^2-3n)T_4 + 6T_5] \end{aligned}$$

where

$$(7.14)\quad \begin{aligned} T_1 &= \sum_{i=1}^{n}\sum_{j=1}^{n}(\tilde{x}_i' C\tilde{a}_j)^4, & T_2 &= \left(\sum_{i=1}^{n}\tilde{a}_i' C' C\tilde{a}_i\right)^2 \\ T_3 &= \sum_{i=1}^{n}(\tilde{a}_i' C' C\tilde{a}_i)^2, & T_4 &= \sum_{i=1}^{n}\left\{\sum_{j=1}^{n}(\tilde{x}_i' C\tilde{a}_j)^2\right\}^2, \\ T_5 &= \sum_{\substack{i=1\\ i\neq j}}^{n}\sum_{j=1}^{n}(\tilde{a}_i' C' C\tilde{a}_j)^2 \end{aligned}$$

and

$$(7.15)\quad \begin{aligned} \tilde{\tilde{x}} &= (X - \underset{\sim}{1}\bar{X}')S_1^{-1/2} = (\tilde{x}_1, \ldots, \tilde{x}_n)' \\ \tilde{A} &= (A - \underset{\sim}{1}\bar{a}') = (\tilde{a}_1, \ldots, \tilde{a}_n)', \qquad \bar{a} = A'\underset{\sim}{1}/n. \end{aligned}$$

By using Lemma 7.3, ξ_4 reduces to

$$(7.16)\quad \xi_4 = \tfrac{2}{3}(n-4)!\int_{\mathscr{G}_1}\{(n^3+n^2)T_1 - (3n^2-3n)T_4\}\tilde{q}^{(4)}(\operatorname{tr} C'C) \times |C'C|^{(n-p-1)/2}\,dC + K_1$$

where K_1 is a constant. To evaluate this integral, let

$$(7.17)\quad \Phi = \int_{\mathscr{G}_1} c_{11}^4\tilde{q}^{(4)}(\operatorname{tr} C'C)|C'C|^{(n-p-1)/2}\,dC,$$

and let $\|b\|^2 = b'b$ for a vector b. Then the following result holds.

Lemma 7.4. *For any two p-vectors α and β,*

$$(7.18)\quad \int_{\mathscr{G}_1}(\alpha' C\beta)^4\tilde{q}^{(4)}(\operatorname{tr} C'C)|C'C|^{(n-p-1)/2}\,dC = \|\alpha\|^4\|\beta\|^4\Phi.$$

Proof. Define $\hat{\alpha} = \alpha/\|\alpha\|$, $\hat{\beta} = \beta/\|\beta\|$ and choose orthogonal $p \times p$ matrices P and Q such that $\hat{\alpha}'$ is the first row of P while $\hat{\beta}$ is the first column of Q. Since the measure $\tilde{q}^{(4)}(\operatorname{tr} C'C)|C'C|^{(n-p-1)/2}$ is orthogonally invariant, making the transformation $C \to PCQ$, the result follows.

This immediately yields

$$(7.19)\quad \int_{\mathscr{G}_1} T_1\tilde{q}^{(4)}(\operatorname{tr} C'C)|C'C|^{(n-p-1)/2}\,dC = \left\{\sum_{i=1}^{n}\|\tilde{x}_i\|^4\right\}\left\{\sum_{i=1}^{n}\|\tilde{a}_i\|^4\right\}\cdot\Phi.$$

To evaluate the other integral $I \equiv \int_{\mathscr{G}_1} T_4 \tilde{q}^{(4)}(\operatorname{tr} C'C)|C'C|^{(n-p-1)/2}\, dC$, we need the following result.

Lemma 7.5.

$$\int_{\mathscr{G}_1} c_{11}^2 c_{12}^2 \tilde{q}^{(4)}(\operatorname{tr} C'C)|C'C|^{(n-p-1)/2}\, dC = \tfrac{1}{3}\Phi.$$

Proof. Let P be any $p \times p$ orthogonal matrix with first column γ. By replacing C in the definition of Φ in (7.17) by CP and, accordingly, c_{11} by the (1,1)th element of CP, namely, $c_1'\underset{\sim}{\gamma}$, where c_1' is the first row of C, we get

(7.20)

$$\Phi = \int_{\mathscr{G}_1} (c_1'\gamma)^4 \tilde{q}^{(4)}(\operatorname{tr} C'C)|C'C|^{(n-p-1)/2}\, dC, \qquad \text{for all} \quad \gamma \quad \text{with} \quad \gamma'\gamma = 1.$$

Take $\gamma' = (2^{-1/2}, 2^{-1/2}, 0, \ldots, 0)$, $(2^{-1/2}, -2^{-1/2}, 0, \ldots 0)$, $(1, 0, \ldots, 0)$, and $(0, 1, \ldots, 0)$ successively, and use the identity $[2^{-1/2}(c_{11} + c_{12})]^4 + [2^{-1/2} \times (c_{11} - c_{12})]^4 = \frac{1}{2}(c_{11}^4 + c_{12}^4 + 6c_{11}^2 c_{12}^2)$ to complete the proof of the lemma.

The integral I is now evaluated as follows.

Lemma 7.6.

$$\int_{\mathscr{G}_1} T_4 \tilde{q}^{(4)}(\operatorname{tr} C'C)|C'C|^{(n-p-1)/2}\, dC$$

$$= \frac{1}{3}\Phi\left(\sum_{i=1}^{n} \|\tilde{x}_i\|^4\right)\left[2\sum_{i=1}^{n}\sum_{j=1}^{n}(\tilde{a}_i'\tilde{a}_j)^2 + \left(\sum_{j=1}^{n}\|\tilde{a}_j\|^2\right)^2\right].$$

Its proof is given in the Appendix. By combining (7.19) and Lemma 7.6, the coefficient of Δ^4, namely ξ_4, is evaluated from (7.16) as

$$\xi_4 = K_1 + \frac{2}{3}(n-4)! \cdot \Phi \cdot \left(\sum_{i=1}^{n}\|\tilde{x}_i\|^4\right) \cdot L(A), \tag{7.21}$$

where

(7.22)

$$L(A) = (n^3 + n^2)\left(\sum_{i=1}^{n}\|\tilde{a}_i\|^4\right) - (n^2 - n)\left[2\sum_{i=1}^{n}\sum_{j=1}^{n}(\tilde{a}_i'\tilde{a}_j)^2 + \left(\sum_{j=1}^{n}\|\tilde{a}_j\|^2\right)^2\right].$$

Going back to (7.8) providing the expression for $H^*(X \mid \Delta)$, the last term in the rhs of (7.8) is $o_X(\Delta^4)$ uniformly in X by Assumption (7.1) and the fact that $\tilde{\tilde{x}}'_\alpha \tilde{\tilde{x}}_\alpha = I_p$, $\forall\, \Gamma_\alpha \in \mathscr{P}$. Hence, the ratio R is finally evaluated as

(7.23) $$R_\Delta(T(X)) = 1 + c_1 \Delta^2 + \{c_2 + \tfrac{2}{3}(n-4)!\Phi L(A) b_{2,p}(X)\}\,\Delta^4 + o_X(\Delta^4)$$

where c_1 and c_2 are constants, and

(7.24) $$b_{2,p}(X) = \sum_{i=1}^{n} \|\tilde{x}_i\|^4 = \sum_{i=1}^{n} \{(X_i - \bar{X})' S^{-1} (X_i - \bar{X})\}^2.$$

This is an expression of (3.94) in Section 3.5 of Chapter 3, as applied to the present problem. An application of the generalized Neyman–Pearson lemma then yields the following final result.

Theorem 7.1. (*Schwagerand Margolin*, 1982; *Sinha*, 1984). *Under the model* (7.1), *where* μ *has the structure* (7.2) *for a specific* A, *and* q *satisfies Assumption* 7.1, *the locally best invariant test of* H: $\Delta = 0$ *versus* K: $\Delta \neq 0$ *rejects* H *for large values of* $\Phi L(A) b_{2,p}(X)$, *where* Φ, $L(A)$ *and* $b_{2,p}(X)$ *are defined in* (7.7), (7.22), *and* (7.24), *respectively*.

Remark 7.1. The test statistic $b_{2,p}(X)$ is the multivariate kurtosis statistic proposed by Mardia(1970). When $p = 1$, $b_{2,p}$ reduces to the usual univariate sample kurtosis b_2.

Remark 7.2. As mentioned in Section 3.2 of Chapter 3, the test statistic $b_{2,p}(X)$ satisfies conditions (i)′ and (ii)′ of Corollary 3.1 of Chapter 3. This establishes the null robustness of the locally optimum test statistic $b_{2,p}(X)$ in $\mathscr{F}_E$ since $\mathscr{F}_E \subset \mathscr{F}_L$.

Remark 7.3. It is stated in the theorem that the locally best test holds for a specific A, depending on the sign of the quantity $L(A)$ defined in (7.22). However, there are some situations under which $L(A) > 0$ for all such configurations of A. This would then imply that the test based on $b_{2,p}\Phi$ is locally best invariant uniformly in such A's. In this context, we mention here one general result without proof. The reader is referred to Schwager and Margolin (1982) for details.

Proposition 7.1. $L(A) > 0$ *whenever the fraction of nonzero rows of* A *is at most* 21%.

This would cover all situations with no more than 21% outliers present in the data.

An interpretation of $L(A)$ for $p = 1$ is interesting. It turns out that $L(A)$ reduces to a multiple of $K_4(a)$, the fourth K-statistic defined in (7.26) below in Remark 7.5.

Remark 7.4. It may be interesting to know what kind of q's satisfy Assumption 7.1. Clearly, for normal scale mixtures, the assumption is satisfied with appropriate moment conditions (Exercise 7.1).

Remark 7.5. Ferguson's (1961) univariate normal result for mean slippage is a special case of the above result when $p = 1$ and $q(u) = (2\pi)^{-n/2}\exp(-u/2)$. In this case, under H: $\Delta = 0$, $x_1, \ldots, x_n$ are iid $N(\mu, \sigma^2)$, and under K: $\Delta \neq 0$, $x_1, \ldots, x_n$ are independent normal with $E(X_i) = \mu + \Delta\sigma a_{v_i}$, $\operatorname{var}(X_i) = \sigma^2$, $i = 1, \ldots, n$, where $a_1, \ldots, a_n$ are known constants and $v = (v_1, \ldots, v_n)$ is an unknown permutation of $(1, \ldots, n)$. The derivation of the locally best invariant test is somewhat simplified because of the assumption $p = 1$. Of course, the assumption of normality of $X_1, \ldots, X_n$ does not simplify the task. The results in Lemma 7.4 and Lemma 7.5 become trivial, and those analogous to Lemma 7.2, Lemma 7.3, and Lemma 7.6 are mentioned below. Recall that $\tilde{\tilde{x}}'_\alpha = (\Gamma_\alpha x - \bar{x}\underset{\sim}{1})'/s$ for $p = 1$ where $s^2 = \Sigma_1^n (x_i - \bar{x})^2$. The proofs given below are different from those previously mentioned. The factor C, which now stands for a scalar, is omitted in what follows. Also note that $A = a$: $n \times 1$. We also write $\tilde{\tilde{x}}_\alpha \equiv \tilde{x}_\alpha$ for notational simplicity, and use the fact that $\tilde{x} = (x - \bar{x}\underset{\sim}{1})'/s$.

Lemma 7.2′. $\Sigma_\alpha(\tilde{x}'_\alpha a)^2 = n\{(n-2)!\}\Sigma_1^n(a_i - \bar{a})^2$.

Proof. $\Sigma_\alpha(\tilde{x}'_\alpha a)^2 = \Sigma_\alpha(\tilde{x}' a_\alpha)^2 = \Sigma_\alpha \tilde{x}' a_\alpha a'_\alpha \tilde{x} = \operatorname{tr} \tilde{x}\tilde{x}'(\Sigma_\alpha a_\alpha a'_\alpha)$.

But, (Exercise 7.2)

$$\sum_\alpha a_\alpha a'_\alpha = n\{(n-2)!\}\left\{\sum_1^n (a_i - \bar{a})^2\right\}(I_n - \underset{\sim}{1}\underset{\sim}{1}'/n) \tag{7.25}$$

and $\operatorname{tr} \tilde{x}\tilde{x}'(I_n - \underset{\sim}{1}\underset{\sim}{1}'/n) = 1$, since $\tilde{x}'\underset{\sim}{1} = 0$ and $\tilde{x}'\tilde{x} = 1$. Thus, the result follows.

Lemma 7.3′.

$$\Sigma_\alpha(\tilde{x}'_\alpha a)^4 = n!\,K_4(a)b_2 + c(n, a),$$

where b_2 is the sample kurtosis $\Sigma_1^n(x_i - \bar{x})^4/\{\Sigma_1^n(x_i - \bar{x})^2\}^2$, $c(n, a)$ is a constant depending on n and a, and $K_4(a)$ is the fourth K-statistic defined by

$$(7.26)\qquad n(n-1)(n-2)(n-3)K_4(a) = (n^3 + n^2)\Sigma(a_i - \bar{a})^4 - 3(n^2 - n)\{\Sigma(a_i - \bar{a})^2\}^2.$$

Note 1: This lemma combines Lemma 7.3 and Lemma 7.6 given in the general case for $p > 1$.

Note 2: Since $\tilde{x}'_\alpha a = \tilde{x}'_\alpha(a - \bar{a}\underset{\sim}{1}) = \tilde{x}'_\alpha\tilde{a} = x'_\alpha\tilde{a}$ for all $\Gamma_\alpha \in \mathscr{P}$ where $\tilde{a} = (a - \bar{a}\underset{\sim}{1})$ and $x_\alpha = \Gamma_\alpha x$, we work below with a and $\tilde{a}$ interchangeably. Also recall that as mentioned in Lemma 7.1, a here stands for a definite permutation $(a_1, \ldots, a_n)$.

Proof. Let $z = (x - \bar{x}\underset{\sim}{1})/s = \tilde{x}$, $z_\alpha = (x_\alpha - \bar{x}\underset{\sim}{1})/s = \tilde{x}_\alpha$ so that $z'\underset{\sim}{1} = z'_\alpha\underset{\sim}{1} = 0$, and $z'z = z'_\alpha z_\alpha = 1$, for all $\Gamma_\alpha \in \mathscr{P}$. Now,

$$(7.27)\qquad \Sigma_\alpha(z'_\alpha a)^4 = \Sigma_\alpha a'z_\alpha z'_\alpha aa'z_\alpha z'_\alpha a = a'(\Sigma_\alpha z_\alpha z'_\alpha aa'z_\alpha z'_\alpha)a,$$

and for each α, the $n \times n$ matrix $z_\alpha z'_\alpha aa'z_\alpha z'_\alpha$ can be written as the product of three matrices of orders $n \times n^2$, $n^2 \times n^2$, $n^2 \times n$, respectively, as

(7.28)

$$z_\alpha z'_\alpha aa'z_\alpha z'_\alpha = \begin{bmatrix} a' & \underset{\sim}{0} & \cdots & \underset{\sim}{0} \\ \underset{\sim}{0} & a' & \ddots & \vdots \\ \vdots & \ddots & \ddots & \underset{\sim}{0} \\ \underset{\sim}{0} & \cdots & \underset{\sim}{0} & a' \end{bmatrix} \begin{bmatrix} z_\alpha z'_\alpha e_1 e'_1 z_\alpha z'_\alpha & \cdots\cdots & z_\alpha z'_\alpha e_1 e'_n z_\alpha z'_\alpha \\ \vdots & & \vdots \\ z_\alpha z'_\alpha e_n e'_1 z_\alpha z'_\alpha & \cdots\cdots & z_\alpha z'_\alpha e_n e'_n z_\alpha z'_\alpha \end{bmatrix} \times \begin{bmatrix} a' & \underset{\sim}{0} & \cdots & \underset{\sim}{0} \\ \underset{\sim}{0} & a' & \ddots & \vdots \\ \vdots & \ddots & \ddots & \underset{\sim}{0} \\ \underset{\sim}{0} & \cdots & \underset{\sim}{0} & a' \end{bmatrix}'$$

where e_k is the vector with 1 at the kth coordinate and 0 elsewhere.

The crux of the problem is to evaluate the sum (over α) of all possible $n^2 \times n^2$ matrices appearing in the middle of (7.28). Straightforward arguments and calculations provide the following:

$$(7.29)\qquad \Sigma_\alpha z_\alpha z'_\alpha e_1 e'_1 z_\alpha z'_\alpha$$

$$= \begin{bmatrix} (n-1)!\Sigma_1^n z_i^4 & (n-2)!\Sigma_{il} z_i^3 z_l & (n-2)!\Sigma_{il} z_i^3 z_l & \cdots & (n-2)!\Sigma_{il} z_i^3 z_l \\ \cdots & (n-2)!\Sigma_{ij} z_i^2 z_j^2 & (n-3)!\Sigma_{ijk} z_i^2 z_j z_k & \cdots & (n-3)!\Sigma_{ijk} z_i^2 z_j z_k \\ \cdots & \cdots\cdots & (n-2)!\Sigma_{ij} z_i^2 z_j^2 & \cdots & (n-3)!\Sigma_{ijk} z_i^2 z_j z_k \\ & & & \ddots & \vdots \\ \cdots & \cdots\cdots & \cdots\cdots\cdots\cdots & \cdots & (n-2)!\Sigma_{ij} z_i^2 z_j^2 \end{bmatrix}$$

$$= \begin{bmatrix} (n-1)!\Sigma z_i^4 & -(n-2)!\Sigma z_i^4 & -(n-2)!\Sigma z_i^4 & \cdots & -(n-2)!\Sigma z_i^4 \\ \cdots & (n-2)!(1-\Sigma z_i^4) & (n-3)!(2\Sigma z_i^4 - 1) & \cdots & (n-3)!(2\Sigma z_i^4 - 1) \\ \cdots & \cdots\cdots & (n-2)!(1-\Sigma z_i^4) & \cdots & (n-3)!(2\Sigma z_i^4 - 1) \\ & & & \ddots & \\ \cdots & \cdots\cdots & \cdots\cdots & \cdots & (n-2)!(1-\Sigma z_i^4) \end{bmatrix}$$

$$= (\Sigma z_i^4)(n-3)! \begin{bmatrix} (n-1)(n-2) & -(n-2) & -(n-2) & \cdots & -(n-2) \\ & -(n-2) & 2 & \cdots & 2 \\ & & -(n-2) & \cdots & 2 \\ & & & \ddots & \\ & & & & -(n-2) \end{bmatrix}$$

$$+ (n-3)! \begin{bmatrix} 0 & 0 & 0 & \cdots & 0 \\ & (n-2) & -1 & \cdots & -1 \\ & & (n-2) & \cdots & -1 \\ & & & \ddots & \vdots \\ & & & & (n-2) \end{bmatrix}$$

and, in general, $\Sigma_\alpha z_\alpha z'_\alpha e_k e'_k z_\alpha z'_\alpha$ is obtained by interchanging the first row, first column of the above matrix with the kth row, kth column. Thus we obtain

(7.30) $$\Sigma_\alpha z_\alpha z'_\alpha e_k e'_k z_\alpha z'_\alpha = (\Sigma z_i^4)(n-3)![2\underset{\sim}{1}\underset{\sim}{1}' + (n^2 - 2n)e_k e'_k - n\,\Delta_k] + D_{n,k},$$

where Δ_k is a matrix with 1 along the diagonal and also in the kth row and kth column and 0 elsewhere, and $D_{n,k}$ is a matrix of constants that need not be computed.

(7.31) $$\Sigma_\alpha z_\alpha z'_\alpha e_1 e'_2 z_\alpha z'_\alpha$$

$$= \begin{vmatrix} (n-2)!\sum_{i\neq j} z_i^3 z_j & (n-2)!\sum_{i\neq j} z_i^2 z_j^2 & (n-3)!\sum_{i\neq j\neq j'} z_i^2 z_j z_{j'} & \rightarrow\cdots & (n-3)!\sum_{i\neq j\neq j'} z_i^2 z_j z_{j'} \\ & (n-2)!\sum_{i\neq j} z_i^3 z_j & (n-3)!\sum_{i\neq j\neq j'} z_i^2 z_j z_{j'} & \rightarrow\cdots & (n-3)!\sum_{i\neq j\neq j'} z_i^2 z_j z_{j'} \\ & & (n-3)!\sum_{i\neq j\neq j'} z_i^2 z_j z_{j'} & \cdots\leftarrow & (n-4)!\sum_{i\neq j\neq 1\neq 1'} z_i z_j z_1 z_{1'} \\ & & \searrow\ddots & & \vdots \\ & & & & (n-3)!\sum_{i\neq j\neq j'} z_i^2 z_j z_{j'} \end{vmatrix}$$

$$= \begin{vmatrix} -(n-2)!\Sigma z_i^4 & (n-2)!(1-\Sigma z_i^4) & (n-3)!(2\Sigma z_i^4 - 1) & \rightarrow\cdots & (n-3)!(2\Sigma z_i^4 - 1) \\ & -(n-2)!\Sigma z_i^4 & (n-3)!(2\Sigma z_i^4 - 1) & \rightarrow\cdots & (n-3)!(2\Sigma z_i^4 - 1) \\ & & (n-3)!(2\Sigma z_i^4 - 1) & \cdots\leftarrow & (n-4)!(3-6\Sigma z_i^4) \\ & & \searrow\ddots & & \downarrow\ \vdots \\ & & & & (n-3)!(2\Sigma z_i^4 - 1) \end{vmatrix}$$

$$= (\Sigma z_i^4)(n-4)! \begin{vmatrix} -(n-2)(n-3) & -(n-2)(n-3) & 2(n-3) & \rightarrow\cdots & 2(n-3) \\ & -(n-2)(n-3) & 2(n-3) & \rightarrow\cdots & 2(n-3) \\ & & 2(n-3) & \cdots\leftarrow & -6 \\ & & & \searrow\ddots & \downarrow\vdots \\ & & & & 2(n-3) \end{vmatrix}$$

$$+ (n-4)! \begin{vmatrix} 0 & (n-2)(n-3) & -(n-3) & \rightarrow\cdots & -(n-3) \\ 0 & & -(n-3) & \rightarrow\cdots & -(n-3) \\ & & -(n-3) & \cdots\leftarrow & 3 \\ & & & \searrow\ddots & \downarrow\vdots \\ & & & & -(n-3) \end{vmatrix}$$

and, in general, $\Sigma_\alpha z_\alpha z'_\alpha e_k e_{k'} z_\alpha z'_\alpha$ is obtained by interchanging the first row, first column of the above matrix with the kth row, kth column, and second row, second column with the k'th row, k'th column.

Thus we can write

$$\begin{aligned} \Sigma_\alpha z_\alpha z'_\alpha e_k e_{k'} z_\alpha z'_\alpha = (\Sigma z_i^4)(n-4)![-6\underset{\sim}{1}\underset{\sim}{1}' + (-n^2+3n) \\ \cdot (e_k e'_k + e_{k'} e'_{k'} + e_k e'_{k'} + e_{k'} e'_k) \\ + 2n\,\Delta_{k,k'}] + E_{n,k,k'}, \end{aligned} \tag{7.32}$$

where $\Delta_{k,k'}$ is a matrix with 1 along the diagonal and also in the kth row, kth column, k'th row, k'th column and 0 elsewhere, and $E_{n,k,k'}$ is a matrix of constants that need not be computed.

By combining (7.28), (7.30), and (7.32), we get

$$\Sigma_\alpha z_\alpha z'_\alpha a a' z_\alpha z'_\alpha = (\Sigma z_i^4)(n-4)!((\delta_{kk'})) + F_a, \tag{7.33}$$

where F_a is a matrix of constants and

$$\begin{aligned} \delta_{kk} &= 2(n-3)(a'\underset{\sim}{1})^2 + (n-3)(n^2-2n)(a'e_k)^2 - n(n-3)a'\Delta_k a, \\ \delta_{kk'} &= -6(a'\underset{\sim}{1})^2 + (-n^2+3n)\{(a'e_k)^2 + (a'e_{k'})^2 + 2(a'e_k)(a'e_{k'})\} \\ &\quad + 2na'\Delta_{k,k'}a, \qquad k \neq k'. \end{aligned} \tag{7.34}$$

Clearly, $a'\Delta_k a = \Sigma a_i^2 + 2a_k \Sigma a_i - 2a_k^2$ for all k, and $a'\Delta_{k,k'}a = (\Sigma a_i)^2 - \Sigma_{i\neq j\neq k\neq k'} a_i a_j$ for all k, k', $k \neq k'$. Recalling Note 2, we can work with $a = a - \bar{a}\underset{\sim}{1}$ in place of a. This immediately gives

$$\begin{aligned} \delta_{kk} &= n^2(n-3)(a_k - \bar{a})^2 - n(n-3)\Sigma(a_k - \bar{a})^2, \\ \delta_{kk'} &= 2n\Sigma(a_i - \bar{a})^2 - (n^2+n)\{(a_k - \bar{a})^2 + (a_{k'} - \bar{a})^2\} \\ &\quad - 2(n^2-n)(a_k - \bar{a})(a_{k'} - \bar{a}), \qquad k \neq k', \end{aligned} \tag{7.35}$$

leading to (see 7.27))

(7.36)
$$(a - \bar{a}1)'((\delta_{kk'}))(a - \bar{a}\underset{\sim}{1}) = (n^3 + n^2)\Sigma(a_i - \bar{a})^4 - 3(n^2 - n)\{\Sigma(a_i - \bar{a})^2\}^2.$$

Since $\Sigma z_i^4 = b_2$, the proof of the lemma is complete.

Note 3: The proof given above is different from Ferguson's and can be adapted to provide a separate and different proof of Lemma 7.3 in the general case from the one given in the Appendix. The reader should note that both proofs are long and quite involved.

7.3. Test for Dispersion Slippage

The model under consideration refers to a data matrix X: $n \times p$ with the distribution

$$(7.37) \quad X \sim \mathscr{F}_E(\underset{\sim}{1}\alpha', \Phi \times \Sigma), \qquad \alpha \in R^p, \qquad \Phi \in \mathscr{S}(n), \qquad \Sigma \in \mathscr{S}(p),$$

where Φ is given in (3.22) of Chapter 3, namely,

$$\Phi = \text{diag}(\exp(\Delta_{a_{v_1}}), \ldots, \exp(\Delta_{a_{v_n}})) \equiv D_\Delta.$$

As mentioned in Sections 3.2 and 3.4 of Chapter 3, this setup with $\Delta = 0$ corresponds to the case of no outlier in the data, whereas a positive value of Δ would suggest presence of outliers. It is assumed that $a_1, \ldots, a_n$ are known nonnegative constants, not all equal and at least one of them positive, but the permutation $(v_1, \ldots, v_n)$ of $(1, \ldots, n)$ is unknown, so that rejection of H: $\Delta = 0$ would not identify the outlying observation vectors.

The model (7.37) given above resembles model (3.5) in Section 3.2 of Chapter 3 except that the mean matrix μ is taken to be of the form $\mu = \underset{\sim}{1}\alpha'$, and the normality there is replaced by the class of elliptically symmetric densities of the form

$$(7.38) \qquad f(X \mid \alpha, \Phi, \Sigma) = |\Phi|^{-p/2}|\Sigma|^{-n/2}q(\text{tr}\,\Sigma^{-1}(X - \underset{\sim}{1}\alpha')'\Phi^{-1}(X - \underset{\sim}{1}\alpha'))$$

for some q: $[0, \infty) \to [0, \infty)$ such that $\int_{R^{np}} q(\text{tr}\,X'X)\,dX = 1$. The distributional form (7.38) appears many times in this book (e.g., in equation (6.2) of Chapter 6).

For the problem of testing H: $\Delta = 0$ versus K: $\Delta > 0$ for unknown $\alpha \in R^p$ and $\Sigma \in \mathscr{S}(p)$ in the model (7.37) or (7.38), the relevant group $\mathscr{G}$ of transformations keeping the problem invariant was mentioned in Section 3.4

of Chapter 3. Briefly,

$$\mathcal{G} = \mathcal{G}_1 \times \mathcal{G}_2 \times \mathcal{G}_3 = \mathrm{GL}(p) \times \mathcal{P} \times R^p,$$

exactly as in the first problem (mean slippage) with the same group action as

$$X \to g(X) = \Gamma X C + \underset{\sim}{1}\eta' \qquad \text{for} \quad g = (C, \Gamma, \eta) \in \mathcal{G}.$$

Theorem 3.10 in Chapter 3 presents a formal expression of the probability ratio R of a maximal invariant π. The corresponding result when α is assumed known and equal to 0 appears in Theorem 3.9 of that chapter. It may be noted that the proofs of these theorems are based on a general approach to nonnull robustness as given in Theorem 3.5 of Chapter 3. We provide below a direct proof of Theorem 3.10, as mentioned there.

Recall (7.4), (7.5), the choice of the invariant measure ν on $\mathcal{G}$, and the fact that due to the invariance of the testing poblem, we may take without loss of generality $\alpha = \underset{\sim}{0}$ and $\Sigma = I_p$. The numerator $H(X \mid \Delta)$ of R is evaluated as

$$(7.39) \qquad H(X \mid \Delta)$$

$$\begin{aligned}
&= \int_{\mathcal{G}} f(g(X) \mid \Delta)\chi(g)\nu(dg) \\
&= |D_\Delta|^{-p/2} \cdot \frac{1}{n!} \sum_\alpha \int_{R^p} \left\{ \int_{\mathcal{G}_1} q[\operatorname{tr}(\Gamma_\alpha X C + \underset{\sim}{1}\eta')' D_\Delta^{-1}(\Gamma_\alpha X C + \underset{\sim}{1}\eta')] \right. \\
&\qquad\qquad \left. \times |C'C|^{(n-p-1)/2}\, dC \right\} d\eta \\
&= |D_\Delta|^{-p/2} \cdot \frac{1}{n!} \sum_\alpha \int_{\mathcal{G}_1} \left\{ \int_{R^p} q[\tau\tilde{\eta}'\tilde{\eta} + \operatorname{tr} C'X'\Gamma_\alpha'(D_\Delta^{-1} - D_\Delta^{-1}\underset{\sim}{1}\underset{\sim}{1}'D_\Delta^{-1}/\tau)\Gamma_\alpha X C]\, d\tilde{\eta} \right\} \\
&\qquad\qquad \times |C'C|^{(n-p-1)/2}\, dC \quad (\tilde{\eta} = \eta + C'X'\Gamma_\alpha' D_\Delta^{-1}/\tau) \\
&= |D_\Delta|^{-p/2} \cdot \frac{\tau^{-p/2}}{n!} \sum_\alpha \int_{\mathcal{G}_1} \tilde{q}(\operatorname{tr} C'X'\Gamma_\alpha'(D_\Delta^{-1} - D_\Delta^{-1}\underset{\sim}{1}\underset{\sim}{1}'D_\Delta^{-1}/\tau)\Gamma_\alpha X C) \\
&\qquad\qquad \times |C'C|^{(n-p-1)/2}\, dC \\
&= |D_\Delta|^{-p/2} \cdot \frac{\tau^{-p/2}}{n!} \sum_\alpha |S_\alpha(\Delta)|^{-(n-1)/2} \cdot \int_{\mathcal{G}_1} \tilde{q}(\operatorname{tr} C'C) |C'C|^{(n-p-1)/2}\, dC,
\end{aligned}$$

where $\tau = \underset{\sim}{1}'D_\Delta^{-1}\underset{\sim}{1}$, $\tilde{q}(u) = \int_{R^p} q(u + x'x)\, dx$, and $S_\alpha(\Delta)$ is defined in (3.80) of Chapter 3, namely

$$S_\alpha(\Delta) = X'\Gamma_\alpha'(D_\Delta^{-1} - D_\Delta^{-1}\underset{\sim}{1}\underset{\sim}{1}'D_\Delta^{-1}/\tau)\Gamma_\alpha X.$$

The denominator of R is precisely $H(X \mid 0)$, given by

$$H(X \mid 0) = \frac{n^{-p/2}}{n!} \sum_{\alpha} |S|^{-(n-1)/2} \int_{\mathscr{G}_1} \tilde{q}(\operatorname{tr} C'C)|C'C|^{(n-p-1)/2}\, dC \tag{7.40}$$

with $S = X'(I_n - \underset{\sim}{1}\underset{\sim}{1}'/n)X = S_\alpha(0)$. Taking the ratio yields

$$R = |D_\Delta|^{-p/2} \cdot \tau^{-p/2} \cdot \frac{n^{p/2}}{n!} \sum_{\alpha} \{|S_\alpha(\Delta)|/|S|\}^{-(n-1)/2}, \tag{7.41}$$

which is (3.79) of Chapter 3. Since R is independent of q, an invariant null-robust test is also nonnull-robust, thereby proving Theorem 3.10. We emphasize that the elements $a_{\nu_1}, \ldots, a_{\nu_n}$ appearing in D_Δ in (7.41) above correspond to a definition permutation, say $(1, \ldots, n)$.

To derive an optimum invariant test, we proceed as in Section 3.5 of Chapter 3. It is easy to show that $H(X \mid \Delta)$ in our problem does not satisfy Assumption 3.3 of Chapter 3, so that a UMPI test does not exist. The derivation of a robust LBI test is routine; just use a Taylor expansion of R around $\Delta = 0$ similar to and much simpler than what we used in Section 7.2 for the mean slippage problem.

The following expansions needed in the sequel are straightforward.

$$\begin{aligned} D_\Delta^{-1} &= \operatorname{diag}(\exp(-\Delta a_1), \ldots, \exp(-\Delta a_n)) \\ &= I_n - \Delta D(a) + \frac{\Delta^2}{2} D(a^2) + o(\Delta^2), \end{aligned} \tag{7.42}$$

where $a^j = (a_1^j, \ldots, a_n^j)$ and $D(a^j) = \operatorname{diag}(a_1^j, \ldots, a_n^j)$. (Exercise 7.3)

$$\tau = n - \Delta \Sigma a_i + \frac{\Delta^2}{2} \Sigma a_i^2 + o(\Delta^2). \qquad \text{(Exercise 7.4)} \tag{7.43}$$

$$\begin{aligned} D_\Delta^{-1}\underset{\sim}{1}\underset{\sim}{1}'D_\Delta^{-1}/\tau = \Big[& \underset{\sim}{1}\underset{\sim}{1}' - \Delta(a\underset{\sim}{1}' + \underset{\sim}{1}a' - \bar{a}\underset{\sim}{1}\underset{\sim}{1}') + \Delta^2 aa' \\ & + \frac{\Delta^2}{2}(a^2\underset{\sim}{1}' + \underset{\sim}{1}a^{2\prime} - \bar{a}^2\underset{\sim}{1}\underset{\sim}{1}') + o(\Delta^2) \Big] \Big/ n \end{aligned} \tag{7.44}$$

where $\bar{a}^2 = \Sigma a_j^2/n$ (Exercise 7.5).

Write $x_\alpha^{*\prime} = (\Gamma_\alpha X - \underset{\sim}{1}\bar{X}')'$ and $\tilde{\tilde{x}}_\alpha' = S^{-1/2}x_\alpha^{*\prime}$. Then

$$\begin{aligned} S_\alpha(\Delta) = {} & S - \Delta x_\alpha^{*\prime} D(a) x_\alpha^* + \frac{\Delta^2}{2} x_\alpha^{*\prime} D(a^2) x_\alpha^* - \Delta^2 x_\alpha^{*\prime} aa' x_\alpha^* \\ & + \text{a remainder term } \delta_x^* \text{ which for every fixed } x \text{ is } o(\Delta^2), \end{aligned} \tag{7.45}$$

implying

$$|S_\alpha(\Delta)|/|S| = \Big| I_p - \Delta \tilde{\tilde{x}}'_\alpha D(a) \tilde{\tilde{x}}_\alpha + \frac{\Delta^2}{2} \tilde{\tilde{x}}'_\alpha (D(a^2) - 2aa') \tilde{\tilde{x}}_\alpha$$

$$+ \text{ a remainder term } \delta_x \text{ which is } o(\Delta^2) \text{ uniformly in } x \Big|$$

(7.46)

$$= 1 - \Delta \operatorname{tr} D(a) \tilde{\tilde{x}}_\alpha \tilde{\tilde{x}}'_\alpha + \frac{\Delta^2}{2} \operatorname{tr}(D(a^2) - 2aa') \tilde{\tilde{x}}_\alpha \tilde{\tilde{x}}'_\alpha$$

$$+ \frac{\Delta^2}{2} \{(\operatorname{tr} D(a) \tilde{\tilde{x}}_\alpha \tilde{\tilde{x}}'_\alpha)^2 - \operatorname{tr}(D(a) \tilde{\tilde{x}}_\alpha \tilde{\tilde{x}}'_\alpha)^2\} + \delta_\alpha(x, \Delta),$$

where $\sup_x \delta_\alpha(x, \Delta) = o(\Delta^2)$, for all $\Gamma_\alpha \in \mathscr{P}$. The second equality in (7.46) is based on the following elementary result whose proof is left as an exercise (Exercise 7.6).

Lemma 7.7.

$$|I_p - \Delta B| = 1 - \Delta \operatorname{tr} B + \frac{\Delta^2}{2} \{(\operatorname{tr} B)^2 - \operatorname{tr} B^2\} + o(\Delta^2).$$

To compute $\Sigma_\alpha \{|S_\alpha(\Delta)|/|S|\}^{-(n-1)/2}$, we also need the following.

(7.47) $$\Sigma_\alpha \operatorname{tr} D(a) \tilde{\tilde{x}}_\alpha \tilde{\tilde{x}}'_\alpha = \Sigma_\alpha \Sigma_i a_{\alpha(i)} (X_i - \bar{X})' S^{-1} (X_i - \bar{X})$$

$$= (n!)(\bar{a}) \Sigma_i (X_i - \bar{X})' S^{-1} (X_i - \bar{X}) = (n!)(\bar{a}) p,$$

using $\Sigma_i (X_i - \bar{X})' S^{-1} (X_i - \bar{X}) = \operatorname{tr} SS^{-1} = p.$

Analogously,

(7.48) $$\Sigma_\alpha \operatorname{tr} D(a^2) \tilde{\tilde{x}}_\alpha \tilde{\tilde{x}}'_\alpha = (n!)(\bar{a}^2) p.$$

(7.49) $$\Sigma_\alpha a' \tilde{\tilde{x}}_\alpha \tilde{\tilde{x}}'_\alpha a = \operatorname{tr}\{S^{-1}(\Sigma_\alpha x^{*\prime}_\alpha a a' x^*_\alpha)\}$$

$$= \operatorname{tr}\{S^{-1}(n(n-2)! \Sigma(a_i - \bar{a})^2 (\Sigma_i (X_i - \bar{X})(X_i - \bar{X})'))\}$$

$$= n(n-2)! p \Sigma(a_i - \bar{a})^2.$$

The second equality above follows easily from Lemma 7.2′.

$$\Sigma_\alpha (\operatorname{tr} D(a) \tilde{\tilde{x}}_\alpha \tilde{\tilde{x}}'_\alpha)^2 = \Sigma_\alpha \{\Sigma_i a_{\alpha(i)} (X_i - \bar{X})' S^{-1} (X_i - \bar{X})\}^2$$

$$= n(n-2)! \Sigma_i \{(X_i - \bar{X})' S^{-1} (X_i - \bar{X})\}^2 \Sigma_i (a_i - \bar{a})^2$$

(7.50)

$$+ (n-2)!((\Sigma a_i)^2 - \Sigma a_i^2) \{\Sigma_i (X_i - \bar{X})' S^{-1} (X_i - \bar{X})\}^2$$

$$= n(n-2)! \{\Sigma(a_i - \bar{a})^2\} b_{2,p}(X) + p^2 (n-2)! \{(\Sigma a_i)^2 - \Sigma a_i^2\}.$$

The second equality above follows from (7.25) of Lemma 7.2′ upon working with $(X_i - \bar{X})'S^{-1}(X_i - \bar{X})$ instead of $(X_i - \bar{X})$.

$$\begin{aligned}
&\Sigma_\alpha \operatorname{tr}(D(a)\tilde{\tilde{x}}_\alpha \tilde{\tilde{x}}_\alpha')^2 \\
&\quad = \Sigma_\alpha \operatorname{tr}\{\Sigma_i a_{\alpha(i)} S^{-1}(X_i - \bar{X})(X_i - \bar{X})'\}^2 \\
&\quad = \operatorname{tr}[n(n-2)!\{\Sigma(a_i - \bar{a})^2\} \\
&\qquad \times \{\Sigma_i S^{-1}(X_i - \bar{X})(X_i - \bar{X})'S^{-1}(X_i - \bar{X})(X_i - \bar{X})'\} \\
(7.51) \quad &\qquad + (n-2)!((\Sigma a_i)^2 - \Sigma a_i^2)\{\Sigma_i S^{-1}(X_i - \bar{X})(X_i - \bar{X})'\}^2] \\
&\quad = n(n-2)!\{\Sigma(a_i - \bar{a})^2\}[\Sigma_i\{(X_i - \bar{X})'S^{-1}(X_i - \bar{X})\}^2] \\
&\qquad + p(n-2)!((\Sigma a_i)^2 - \Sigma a_i^2) \\
&\quad = n(n-2)!\{\Sigma(a_i - \bar{a})^2\} b_{2,p}(X) + p(n-2)!((\Sigma a_i)^2 - \Sigma a_i^2).
\end{aligned}$$

The second equality above follows from (7.25) when applied component-wise. Alternatively, denoting $\theta_{ij} = (X_i - \bar{X})'S^{-1}(X_j - \bar{X})$, $i, j = 1, \ldots, n$, it follows that

$$(7.52) \qquad \sum_\alpha \operatorname{tr}(D(a)\tilde{\tilde{x}}_\alpha \tilde{\tilde{x}}_\alpha')^2 = \sum_i \theta_{ii}^2 \left(\sum_\alpha a_{\alpha(i)}^2 \right) + \sum_{i \neq j} \theta_{ij}^2 \left(\sum_\alpha a_{\alpha(i)} a_{\alpha(j)} \right).$$

But

$$\begin{aligned}
(7.53) \qquad &\Sigma_\alpha a_{\alpha(i)}^2 = (n-1)!\Sigma_i a_i^2 \qquad \text{and} \\
&\Sigma_\alpha a_{\alpha(i)} a_{\alpha(j)} = (n-2)![(\Sigma a_i)^2 - \Sigma a_i^2], \quad i \neq j.
\end{aligned}$$

Substituting (7.53) in (7.52) and noting that $\Sigma_i \theta_{ii}^2 = b_{2,p}(X)$, $\Sigma_{i \neq j} \theta_{ij}^2 = p - b_{2,p}(X)$, the result follows.

We are now ready to obtain a local expansion of R around $\Delta = 0$. Note that the expression for R is given in (7.41). Using the above computations (7.42) through (7.51), a straightforward computation provides the following expansion of R.

$$(7.54) \qquad R = 1 + \Delta^2 \left\{ d_1 + \frac{n+1}{8} \Sigma(a_i - \bar{a})^2 \cdot b_{2,p}(X) \right\} + o(\Delta^2),$$

where d_1 is a constant and the term $o(\Delta^2)$ is uniform in X.

This is an expression of (3.94) in Section 3.5 of Chapter 3 as applied to the present problem. Therefore, we have proved the following result.

Theorem 7.2. (*Das and Sinha*, 1986). *Under the model* (7.37) *with* $\Phi = D_\Delta$, *the locally best invariant test of* H: $\Delta = 0$ *versus* K: $\Delta > 0$ *rejects* H *for a large value of* $b_{2,p}(X)$, *whatever* $a_1, \ldots, a_n$. *The test is null, nonnull, and optimality-robust.*

Remark 7.6. The null robustness of $b_{2,p}(X)$ is mentioned in Remark 7.2 of Section 7.2.

Remark 7.7. The local optimality (subject to invariance) of the same test statistic $b_{2,p}(X)$ for both mean and dispersion slippage is indeed rewarding, as it provides simultaneous protection against two different types of shifts from a homogeneous sample.

Remark 7.8. A formal direct proof of Theorem 3.9 in Chapter 3 when $\alpha = 0$ can be given along the same lines used to derive (7.41).

Remark 7.9. The univariate normal theory result of Ferguson (1961) for variance slippage follows as a special case of our general result given above upon taking $p = 1$ and $q(u) = (2\pi)^{-n/2}\exp(-u/2)$. Of course, choosing $q(u)$ in this way does not have any advantage because of non-null robustness of all null-robust invariant tests. That is to say, the choice of q does not play any role because R in (7.41) is independent of q. However, there are some simplifications in some of the formulae (7.45)–(7.51), which are needed to derive the LBI test. We emphasize that the constant $\frac{n+1}{8}$ appearing in (7.54) prevails for all p.

Remark 7.10. The distributional assumption about X, namely (7.38), can be somewhat broadened to the densities of the form

$$(7.55)\qquad f(X \mid \alpha, \Phi, \Sigma) = |\Phi|^{-p/2}|\Sigma|^{-n/2}q(\Sigma^{-1}(X - \underset{\sim}{1}\alpha')'\Phi^{-1}(X - \underset{\sim}{1}\alpha')),$$

where $q\colon M(p) \to [0, \infty)$, $M(p)$ is the set of $p \times p$ matrices of the form VB for $V \in \mathscr{S}(p)$, $B \in \bar{\mathscr{S}}(p)$ (the set of $p \times p$ nonnegative definite matrices), and q satisfying

$$(7.56)\qquad \int_{R^{n\times p}} q(X'X)\,dX = 1, \qquad q(BL) = q(LB), \qquad \forall B \in \mathscr{S}(p), \qquad \forall L \in \mathrm{GL}(p).$$

The conclusion is that for testing H: $\Delta = 0$ versus K: $\Delta > 0$ in the model (7.55) with $\alpha \in R^p$ and $\Sigma \in \mathscr{S}(p)$ unknown, the LBI test rejects H for a large value of

$b_{2,p}(X)$, whatever $a_1, \ldots, a_n$, and with q satisfying (7.56). This is because the expression for R given in (7.41) remains unchanged (Exercise 7.7). The reader may note that this extention is similar to the one for the MANOVA problem discussed in Section 5.3 of Chapter 5, except that here we do not need convexity of q.

Appendix

Proof of Lemma 7.2. Let $B = AC'$. Note that $\tilde{\tilde{x}}'_\alpha = S^{-1/2}(\Gamma_\alpha X - \underset{\sim}{1}\bar{X}')'$ satisfies $\tilde{\tilde{x}}'_\alpha \underset{\sim}{1} = \underset{\sim}{0}$, $\tilde{\tilde{x}}'_\alpha \tilde{\tilde{x}}_\alpha = I_p$, $\forall \Gamma_\alpha \in \mathscr{P}$. By using these properties, it is easy to show that (Exercise 7.8)

$$\text{(A.7.1)} \qquad \Sigma_\alpha(\operatorname{tr} B\tilde{\tilde{x}}'_\alpha)^2 = n(n-2)!(\operatorname{tr} BB') - (n-2)!(\underset{\sim}{1}'BB'\underset{\sim}{1}).$$

By taking $B = AC'$, the lemma follows.

Proof of Lemma 7.3. Since $\tilde{\tilde{x}}'_\alpha \underset{\sim}{1} = \underset{\sim}{0}$, $\tilde{A}'_\alpha \underset{\sim}{1} = \underset{\sim}{0}$, $\forall \Gamma_\alpha \in \mathscr{P}$ where $\tilde{A}_\alpha = (\Gamma_\alpha A - \underset{\sim}{1}\bar{a}')$, we can work with $(\tilde{\tilde{x}}, A)$ or $(\tilde{\tilde{x}}, \tilde{A})$ interchangeably. Write

$$\text{(A.7.2)} \qquad \operatorname{tr} C'\tilde{\tilde{x}}'_\alpha A = \operatorname{tr} C'\tilde{\tilde{x}}'_\alpha \tilde{A} = \operatorname{tr} \tilde{A}C'\tilde{\tilde{x}}'_\alpha = \operatorname{tr} \tilde{\tilde{x}}_\alpha C\tilde{A}' = \sum_{i=1}^n \tilde{x}'_{\alpha(i)} C\tilde{a}_i,$$

where $\tilde{\tilde{x}}_\alpha = (\tilde{x}_{\alpha(1)}, \ldots, \tilde{x}_{\alpha(n)})'$ and $\alpha = (\alpha(1), \ldots, \alpha(n))$ denotes a permutation of $(1, \ldots, n)$. Clearly,

$$\text{(A.7.3)} \qquad \sum_\alpha \left\{ \sum_{i=1}^n \tilde{x}'_{\alpha(i)} C\tilde{a}_i \right\}^4 = \sum_\alpha \left\{ \sum_{i=1}^n \tilde{x}'_i C\tilde{a}_{\alpha(i)} \right\}^4.$$

Denote $y_{\alpha,i} = \tilde{x}'_i C\tilde{a}_{\alpha(i)}$. By the multinomial theorem, we can write

$$\text{(A.7.4)} \qquad \sum_\alpha \left(\sum_{i=1}^n y_{\alpha,i} \right)^4 = \sum_\alpha \Bigg[\sum_i y^4_{\alpha,i} + 4 \sum_{i \neq j} y^3_{\alpha,i} y_{\alpha,j} + 3 \sum_{i \neq j} y^2_{\alpha,i} y^2_{\alpha,j} + 6 \sum_{i \neq j \neq k} y^2_{\alpha,i} y_{\alpha,j} y_{\alpha,k} + \sum_{i \neq i' \neq j \neq j'} y_{\alpha,i} y_{\alpha,j} y_{\alpha,i'} y_{\alpha,j'} \Bigg].$$

By using $\Sigma_i \tilde{x}_i = \underset{\sim}{0}$, $\Sigma_i \tilde{x}_i \tilde{x}'_i = I_p$ and $\Sigma_i \tilde{a}_{\alpha(i)} = 0$, each of the terms on the rhs of (A.7.4) can be expressed in terms of the T_i's given in (7.14). For example,

$$\text{(A.7.5)} \qquad \sum_\alpha \left(\sum_i y^4_{\alpha,i} \right) = \sum_\alpha \left\{ \sum_i (\tilde{x}'_i C\tilde{a}_{\alpha(i)})^4 \right\} = (n-1)! \sum_{i,j} (\tilde{x}'_i C\tilde{a}_j)^4 = (n-1)!\, T_1;$$

$$\sum_{\alpha}\left(\sum_{i\neq j} y_{\alpha,i}^3 y_{\alpha,j}\right) = \sum_{\alpha}\sum_{i\neq j}(\tilde{x}_i' C\tilde{a}_{\alpha(i)})^3(\tilde{x}_j' C\tilde{a}_{\alpha(j)})$$

$$\text{(A.7.6)} \qquad = (n-2)!\sum_{i\neq j}\sum_{i'\neq j'}(\tilde{x}_i' C\tilde{a}_{i'})^3(\tilde{x}_j' C\tilde{a}_{j'})$$

$$= (n-2)!\sum_{i,i'}(\tilde{x}_i' C\tilde{a}_{i'})^3\left[\sum_{j\neq i}\tilde{x}_j' C\left(\sum_{j'\neq i'}\tilde{a}_{j'}\right)\right]$$

$$= (n-2)!\sum_{i,i'}(\tilde{x}_i' C\tilde{a}_{i'})^4 = (n-2)!\,T_1.$$

Analogously, after some tedious but similar calculations, we get the other terms as (Exercise 7.9)

$$\text{(A.7.7)} \qquad \sum_{i\neq j} y_{\alpha,i}^2 y_{\alpha,j}^2 = (n-2)!(T_1 + T_2 - T_3 - T_4),$$

$$\text{(A.7.8)} \qquad \sum_{i\neq j\neq k} y_{\alpha,i}^2 y_{\alpha,j} y_{\alpha,k} = (n-3)!(4T_1 + T_2 - 2T_3 - 2T_4),$$

$$\text{(A.7.9)} \qquad \sum_{i\neq i'\neq j\neq j'} y_{\alpha,i} y_{\alpha,i'} y_{\alpha,j} y_{\alpha,j'} = (n-4)!(36T_1 + 3T_2 - 12T_3 - 18T_4 + 6T_5).$$

By combining (A.7.2) through (A.7.9), the lemma follows.

Proof of Lemma 7.6. Recall that $T_4 = \Sigma_i\{\Sigma_j(\tilde{x}_i' C\tilde{a}_j)^2\}^2$. For any i, we can write

$$\sum_j(\tilde{x}_i' C\tilde{a}_j)^2 = \sum_j \tilde{x}_i' C\tilde{a}_j\tilde{a}_j' C'\tilde{x}_i = \tilde{x}_i' C\left(\sum_j \tilde{a}_j\tilde{a}_j'\right)C'\tilde{x}_i$$

and make the transformation $C \to C^* = PC$ where P is a $p \times p$ orthogonal matrix with the first row as $\tilde{x}_i'/\|\tilde{x}_i\|$. It then follows that

$$\text{(A.7.10)} \qquad \int_{\mathscr{G}_1} T_4\tilde{q}^{(4)}(\operatorname{tr} C'C)|C'C|^{(n-p-1)/2}\,dC = \left(\sum_i\|\tilde{x}_i\|^4\right)\cdot\psi,$$

where

$$\text{(A.7.11)} \qquad \psi = \int_{\mathscr{G}_1}\left\{c_1'\left(\sum_j\tilde{a}_j\tilde{a}_j'\right)c_1\right\}^2\tilde{q}^{(4)}(\operatorname{tr} C'C)|C'C|^{(n-p-1)/2}\,dC,$$

and c_1' denotes the first row of C. To evaluate ψ, let V be $p \times p$ orthogonal such that $V(\Sigma_j\tilde{a}_j\tilde{a}_j')V' = \operatorname{diag}(\lambda_1,\ldots,\lambda_p)$, where $\lambda_1,\ldots,\lambda_p$ are the eigen values of $(\Sigma_j\tilde{a}_j\tilde{a}_j')$. By making the transformation $C \to C^* = VC$, it follows easily from the definition of Φ given in (7.17) and Lemma 7.5 that

$$\text{(A.7.12)} \qquad \psi = \left(\sum_i \lambda_i^2\right)\Phi + \left(\sum_{i \neq j} \lambda_i \lambda_j\right)\tfrac{1}{3}\Phi = \tfrac{1}{3}\Phi\left[2\sum_i \lambda_i^2 + (\Sigma\lambda_i)^2\right].$$

Finally, we note that $\Sigma_i \lambda_i = \text{tr}(\Sigma_j \tilde{a}_j \tilde{a}_j') = \Sigma_j \|\tilde{a}_j\|^2$ and $\Sigma\lambda_i^2 = \text{tr}(\Sigma_j \tilde{a}_j \tilde{a}_j')^2 = \Sigma_{j,k}(\tilde{a}_j'\tilde{a}_k)^2$. This completes the proof.

Exercises

7.1. Show that if

$$f(x \mid \mu, \Sigma) = \frac{|\Sigma|^{-n/2}}{(2\pi)^{np/2}} \int_0^\infty \exp[-a\,\text{tr}\,\Sigma^{-1}(x - \mu)'(x - \mu)]a^{np/2}\,dF(a)$$

for some cdf $F(a)$ satisfying $\int_0^\infty dF(a) = 1$, then Assumption 7.1 is satisfied whenever $\int_0^\infty a^{i-2}\,dF(a) < \infty$ for $i = 1, 2, 3, 4$.

7.2. Prove (7.25). [Hint: Note that the element in the (1, 1) position on the left-hand side is $(n-1)!\Sigma a_i^2$, and that in the (1, 2) position, it is $(n-2)!\Sigma_{i \neq j} a_i a_j$.]

7.3. Prove (7.42). [Hint: Use $\exp(-\Delta a) = 1 - \Delta a + \frac{a}{2}\Delta^2 + o(\Delta^2)$.]

7.4. Prove (7.43). [Hint: Use (7.42) and the definition of τ.]

7.5. Prove (7.44). [Hint: Use (7.42) and (7.43).]

7.6. Prove Lemma 7.7. [Hint: First prove it for B diagonal.]

7.7. Prove that the expression for R in (7.41) remains unchanged when the underlying f satisfies (7.55).

7.8. Prove (A.7.1). [Hint: Expand $(\text{tr}\,B\tilde{\tilde{x}}_\alpha')^2$ and use Exercise (7.2).]

7.9. Prove (A.7.7), (A.7.8), and (A.7.9). [Hint: Proceed as in (A.7.6). You may also use the paper by Schwager and Margolin(1982).]

Chapter 8 Two-Population Problems

8.1. Introduction

So far throughout the book, we have considered different aspects of robustness of various multivariate (and sometimes univariate) tests on the basis of a broad class of nonnormal distributions. However, as is discussed in Chapter 1, the framework for nonnormality we have adopted so far is built on the sacrifice of independence. For example, in Chapters 5, 6, and 7, it has been generally assumed that a data matrix $X: n \times p$ follows a distribution with pdf of the form

$$|\Phi|^{-p/2}|\Sigma|^{-n/2}q(\operatorname{tr}\Sigma^{-1}(X - M)'\Phi^{-1}(X - M)). \tag{8.1}$$

Of course, suitable structures on M, Σ, and Φ are assumed in different contexts. But even if $\Phi = I_n$ holds, which indeed it does most often, the different rows of X are not independent. Therefore, in dealing with independent populations, this approach is not quite natural.

In this chapter, we consider some testing problems involving the equality of location and scale parameters of two independent univariate populations and establish the optimality robustness of some familiar tests. Below, the following two cases are considered separately:

(1) Optimality robustness of some tests derived under normal distribution.

(2) Optimality robustness of some tests derived under exponential distribution.

Case (1) will be called the nonnormal case, whereas (2) will be called the nonexponential case. In the case of two independent nonnormal populations, it is shown in Section 8.2 that for testing the equality of two location parameters without any scale parameter, the test based on the difference between the two sample means is conditional UMPI, conditionally given two ancillary statistics, under certain conditions. In the case of two independent nonexponential populations in Section 8.3, the same property is shown to hold for the test based on the difference between two minimums. However, these tests are not unconditional UMPI unless the two populations are both normal or both exponential, or unless there is only one observation from each population. On the other hand, for testing the equality of two scale parameters with or without location parameters, the standard normal-theory or exponential-theory F-test is still optimal for suitable deviations from normality or exponentiality. This is explained in Sections 8.4 and 8.5. The null robustness of the optimal tests does not hold, though, in all the cases, mainly because of the presence of independence.

8.2. Test of Equality of Two Location Parameters—Nonnormal Case

Let $x^*: (n_1 + 1) \times 1$ and $y^*: (n_2 + 1) \times 1$ be independent random vectors with pdf's (with respect to the Lebesgue measure)

$$
\begin{aligned}
f_{\theta_1}(x^*) &= q_1[(x^* - \theta_1 e_1)'(x^* - \theta_1 e_1)], \\
f_{\theta_2}(y^*) &= q_2[(y^* - \theta_2 e_2)'(y^* - \theta_2 e)],
\end{aligned}
\tag{8.2}
$$

where $x^* \in R^{n_1+1}$, $y^* \in R^{n_2+1}$, $e_1 = (1, \ldots, 1)' \in R^{n_1+1}$, $e_2 = (1, \ldots, 1)' \in R^{n_2+1}$, θ_1, $\theta_2 \in R$ and q_1 and q_2 are pdf's over R^{n_1+1} and R^{n_2+1}, respectively. Making a suitable orthogonal transformation, without any loss of generality, we can write $x^* = (x, x_1')'$, $y^* = (y, y_1')'$, where $x_1 \in R^{n_1}$, $y_1 \in R^{n_2}$, and assume that the pdf's are

$$
f_{\theta_1}(x^*) = q_1[(x - \theta_1)^2 + x_1' x_1], \qquad f_{\theta_2}(y^*) = q_2[(y - \theta_2)^2 + y_1' y_1]. \tag{8.3}
$$

This formulation is different from what has been assumed in the previous chapters, namely the formulation that (x^*, y^*) jointly has a pdf of the form

$$
q[(x - \theta_1)^2 + (y - \theta_2)^2 + x_1' x_1 + y_1' y_1], \qquad \text{for some} \quad q,
$$

thus making x^* and y^* dependent. In the model (8.3), we consider the problem of testing $\mathrm{H}_0: \theta_1 = \theta_2$ versus $\mathrm{H}_1: \theta_1 > \theta_2$. This problem remains invariant under the group G of transformations, $G = G_1 \times G_2$, where G_1 is the group of translations: $x \to x + c$, $y \to y + c$, $\theta_1 \to \theta_1 + c$, $\theta_2 \to \theta_2 + c$, $-\infty < c < \infty$, and G_2 is the group of orthogonal transformations: $\mathcal{O}(n_1) \times \mathcal{O}(n_2)$ acting on x_1 and y_1 as: $x_1 \to \Gamma_1 x_1$, $y_1 \to \Gamma_2 y_1$, $\Gamma_i \in \mathcal{O}(n_i)$, $i = 1, 2$. As a left-invariant measure ν on G, we take $\nu(dg) = dc\nu_1(d\Gamma_1)\nu_2(d\Gamma_2)$ where $\nu_i(d\Gamma_i)$ is the invariant probability measure on $\mathcal{O}(n_i)$ $(i = 1, 2)$. Clearly $T = (x - y, x_1'x_1, y_1'y_1)$ is a maximal invariant. By using Wijsman's representation theorem (see Chapter 2), the ratio of the pdf's of T under nonnull to null is obtained as

$$
\begin{aligned}
(8.4)\qquad R &= dP^T_{H_1}/dP^T_{H_0} \\
&= \frac{\int_{-\infty}^{\infty} q[(x + c - \theta_1)^2 + x_1'x_1]q_2[(y + c - \theta_2)^2 + y_1'y_1]\,dc}{\int_{-\infty}^{\infty} q_1[(x + c - \theta_1)^2 + x_1'x_1]q_2[(y + c - \theta_1)^2 + y_1'y_1]\,dc} \\
&= \frac{\int_{-\infty}^{\infty} q_1[c^2 + x_1'x_1]q_2[(c - z + \delta)^2 + y_1'y_1]\,dc}{\int_{-\infty}^{\infty} q_1[c^2 + x_1'x_1]q_2[(c - z)^2 + y_1'y_1]\,dc} \\
&= \frac{\int_{-\infty}^{\infty} q_1[(c - \delta)^2 + x_1'x_1]q_2[(c - z)^2 + y_1'y_1]\,dc}{\int_{-\infty}^{\infty} q_1[c^2 + x_1'x_1]q_2[(c - z)^2 + y_1'y_1]\,dc}.
\end{aligned}
$$

where $z = x - y$, $\delta = \theta_1 - \theta_2$. To derive an optimum invariant test based on R, note that, conditionally given the two ancillary statistics $x_1'x_1$ and $y_1'y_1$, the ratio of the conditional pdf's of $(x - y)$ under nonnull to null is precisely R. However, from (8.4), R can be expressed as

$$(8.5)\qquad R = E\{q_1[(c - \delta)^2 + x_1'x_1]/q_1[c^2 + x_1'x_1]\},$$

where E stands for expectation with respect to c having the pdf

$$(8.6)\qquad q_1[c^2 + x_1'x_1]q_2[(c - z)^2 + y_1'y_1]\Big/ \int_{-\infty}^{\infty} q_1[c^2 + x_1'x_1]q_2[(c - z)^2 + y_1'y_1]\,dc.$$

We now make the following assumption.

Assumption 8.1. $q_1, q_2 \in \mathcal{Q}$, where

$$\mathcal{Q} = \left\{ q \colon \frac{q[(c-a)^2 + t]}{q[c^2 + t]} \quad \text{is nowhere nondecreasing in} \quad c \right.$$

$$\left. \text{for any} \quad a > 0, \quad t > 0 \right\}.$$

It then follows that the family of distributions of c in (8.6), generated by the parameter z, keeping $x_1'x_1$ and $y_1'y_1$ fixed, has monotone likelihood ratio (MLR) property in c by the assumption $q_2 \in \mathcal{Q}$, and hence R is nondecreasing in z for $\delta > 0$ by the assumption $q_1 \in \mathcal{Q}$ (Lehmann, 1959, p. 74). We have, therefore, proved the following (Kariya and Sinha, 1987).

Theorem 8.1. *For testing* $H_0\colon \theta_1 = \theta_2$ *versus* $H_1\colon \theta_1 > \theta_2$ *in the model* (8.3), *the test with a critical region* $z > k$ *is UMPI in the class of conditional level* α *tests, conditionally given the two ancillary statistics* $x_1'x_1$ *and* $y_1'y_1$, *whenever* $q_1, q_2 \in \mathcal{Q}$.

Remark 8.1. It is clear from (8.4) that when both q_1 and q_2 are normal, $x_1'x_1$ and $y_1'y_1$ can be ignored, and the above test is unconditionally UMPI. For nonnormal q's, sufficiency-invariance reduces $(x, y, x_1'x_1, y_1'y_1)$ to $(x - y, x_1'x_1, y_1'y_1)$, and the above conditional argument is necessary unless $n_1 = n_2 = 0$. This means the test is UMPI when there is only one observation from each population and $q_1, q_2 \in \mathcal{Q}$.

Remark 8.2. If $q \in \mathcal{Q}$ is differentiable, $\mathcal{Q}$ is equivalent to

$$\tilde{\mathcal{Q}} = \left\{ q\colon \left(\frac{cq'(c^2 + t)}{q(c^2 + t)} \right) \downarrow c, \qquad \forall t \geq 0 \right\}. \quad \text{(Exercise 8.1)} \tag{8.7}$$

An example of a $q \in \tilde{\mathcal{Q}}$ is provided by $q(u) \propto e^{-u^r}$, $r \geq \frac{1}{2}$. It may be remarked that a t-density $\notin \mathcal{Q}$.

Remark 8.3. The reader may verify that an attempt to derive a UMPI or even an LBI test when q_1 and q_2 involve an unknown common scale parameter is not successful unless q_1 and q_2 are normal. It is conjectured that the familiar Fisher t-test for this problem based on $(x - y)/(x_1'x_1 + y_1'y_1)^{1/2}$ is optimum only for normal densities.

Remark 8.4. The null pdf of z is given by

$$dP^z_{H_0}/dz = \int_{-\infty}^{\infty} \tilde{q}_1(c^2)\tilde{q}_2((c-z)^2)\, dc, \tag{8.8}$$

where

$$\tilde{q}_1(c^2) = \int_{R^{n_1}} q_1(c^2 + x_1'x_1)\,dx_1,\ \tilde{q}_2((c-z)^2) = \int_{R^{n_1}} q_2((c-z)^2 + y_1'y_1)\,dy_1.$$

Thus the null robustness of the test does not hold.

8.3. Test of Equality of Two Location Parameters—Nonexponential Case

Let $x: n_1 \times 1$ and $y: n_2 \times 1$ be independent random vectors with pdf's (with respect to the Lebesgue measure)

$$f_{\theta_1}(x) = q_1\left[\sum_1^{n_1}(x_i - \theta_1)\right], \qquad f_{\theta_2}(y) = q_2\left[\sum_1^{n_2}(y_i - \theta_2)\right] \tag{8.9}$$

$$x_i \geq \theta_1, \qquad y_i \geq \theta_2, \qquad x \in R^{n_1}, \qquad y \in R^{n_2}, \qquad \theta_1, \theta_2 \in R.$$

The problem of testing $H_0: \theta_1 = \theta_2$ versus $H_1: \theta_1 > \theta_2$ remains invariant under the group G of translations acting as $x_i \to x_i + c$, $y_i \to y_i + c$, $i = 1, \ldots, n$, $\theta_1 \to \theta_1 + c$, $\theta_2 \to \theta_2 + c$, $-\infty < c < \infty$. Let $x_{(1)} = \min x_i$, $y_{(1)} = \min y_i$, $t_1 = \Sigma_1^{n_1}(x_i - x_{(1)})$, $t_2 = \Sigma_1^{n_2}(y_i - y_{(1)})$. By applying Wijsman's theorem (Chapter 2), the ratio R of the pdf's of a maximal invariant $T = (x_{(1)} - y_{(1)}, t_1, t_2)$ under nonnull to null is obtained as

$$\begin{aligned}
R = \frac{dP^T_{H_1}}{dP^T_{H_0}} &= \frac{\int_B q_1\left[\sum_1^{n_1}(x_i + c - \theta_1)\right] q_2\left[\sum_1^{n_2}(y_i + c - \theta_2)\right] dc}{\int_A q_1\left[\sum_1^{n_1}(x_i + c - \theta_1)\right] q_2\left[\sum_1^{n_2}(y_i + c - \theta_1)\right] dc} \\
&= \frac{\int_B q_1[t_1 + n_1(x_{(1)} + c - \theta_1)]q_2[t_2 + n_2(y_{(1)} + c - \theta_2)]\,dc}{\int_A q_1[t_1 + n_1(x_{(1)} + c - \theta_1)]q_2[t_2 + n_2(y_{(1)} + c - \theta_1)]\,dc} \\
&= \frac{\int_{c \geq \max(0, z-\delta)} q_1[t_1 + c]q_2\left[t_2 + \frac{n_2}{n_1}(c - z + \delta)\right] dc}{\int_{c \geq \max(0, z)} q_1[t_1 + c]q_2\left[t_2 + \frac{n_2}{n_1}(c - z)\right] dc} \\
&= \frac{\int_{c \geq \max(\delta, z)} q_1[t_1 + c - \delta]q_2\left[t_2 + \frac{n_2}{n_1}(c - z)\right] dc}{\int_{c \geq \max(0, z)} q_1[t_1 + c]q_2\left[t_2 + \frac{n_2}{n_1}(c - z)\right] dc},
\end{aligned} \tag{8.10}$$

where $z = x_{(x)} - y_{(1)}$, $\delta = \theta_1 - \theta_2$, $A = \{c: c > \max(\theta_1 - x_{(1)}, \theta_1 - y_{(1)})\}$, and $B = \{c: c > \max(\theta_1 - x_{(1)}, \theta_2 - y_{(1)})\}$. To derive an optimum invariant test based on R, note that, conditionally given the two ancillary statistics t_1 and t_2, the ratio of the conditional pdf's of z under nonnull to null is precisely R. Moreover, from (8.10), R can be expressed as

$$R = E\{q_1[(t_1 + c - \delta)]/q_1[t_1 + c]\} \cdot I_{[c>\delta]}, \tag{8.11}$$

where I is the indicator function, and E stands for expectation with respect to c having the pdf (for $c > \max(0, z)$)

(8.12)

$$q_1[t_1 + c]q_2\left[t_2 + \frac{n_2}{n_1}(c - z)\right] \Big/ \int_{c>\max(0,z)} q_1[t_1 + c]q_2\left[t_2 + \frac{n_2}{n_1}(c - z)\right] dc.$$

We now make the following assumption.

Assumption 8.2. $q_1, q_2 \in \mathcal{Q} = \{q: q(t + c - a)/q(t + c)$ nondecreasing in c for $c > a, t \geq 0\}$.

It then follows that the family of distributions of c in (8.12), generated by z for fixed t_1, t_2, has MLR property in c by the assumption $q_2 \in \mathcal{Q}$, and hence R is nondecreasing in z for $\delta > 0$ by the assumption $q_1 \in \mathcal{Q}$ (Lehmann, 1959, p. 74). This proves the following result (Kariya and Sinha, 1987).

Theorem 8.2. *For testing H_0: $\theta_1 = \theta_2$ versus H_1: $\theta_1 > \theta_2$ in the model (8.9), the test with the critical region $z > k$ is UMPI in the class of conditional level α tests, conditionally given the two ancillary statistics t_1 and t_2, whenever $q_1, q_2 \in \mathcal{Q}$.*

Remark 8.5. It is clear from (8.10) that when both q_1 and q_2 are exponential, t_1 and t_2 can be ignored and the above test is unconditionally UMPI. For nonexponential q's, sufficiency-invariance reduces $(x_{(1)}, y_{(1)}, t_1, t_2)$ to $(x_{(1)} - y_{(1)}, t_1, t_2)$, and the above conditional argument is necessary unless $n_1 = n_2 = 1$, in which case t_1 and t_2 become vacuous.

Remark 8.6. If $q \in \mathcal{Q}$ is differentiable, $\mathcal{Q}$ is equivalent to

$$\tilde{\mathcal{Q}} = \{q: cq'(t + c)/q(t + c) \quad \text{is nonincreasing in} \quad c, \quad t \geq 0\}. \qquad \text{(Exercise 8.2)}$$

An example of a $q \in \tilde{\mathcal{Q}}$ is provided by $q(u) \propto e^{-u^r}, r > 0$.

Remark 8.7. The null pdf of z is given by

$$dP^z_{\mathrm{H}_0}/dz = \int_{c > \max(0,z)} \tilde{q}_1[c]\tilde{q}_2[c - z]\, dc,$$

where

$$\tilde{q}_1(c) = \int_0^\infty q_1[c + t_1]t_1^{(n_1-1)/2}\, dt_1,$$

$$\tilde{q}_2(c - z) = \int_0^\infty q_2[c - z + t_2]t_2^{(n_2-1)/2}\, dt_2.$$

Thus the null robustness of the test does not hold.

8.4. Test of Equality of Two Scale Parameters—Nonnormal Case

Let x: $n_1 \times 1$ and y: $n_2 \times 1$ be independent random vectors with pdf's (with respect to the Lebesgue measure)

$$\text{(8.13)} \qquad \begin{aligned} f_{\mu_1,\sigma_1}(x) &= \sigma_1^{-n_1}q_1((x - \mu_1\underset{\sim}{1}_1)'(x - \mu_1\underset{\sim}{1}_1)/\sigma_1^2), \\ f_{\mu_2,\sigma_2}(y) &= \sigma_2^{-n_2}q_2((y - \mu_2\underset{\sim}{1}_2)'(y - \mu_2\underset{\sim}{1}_2)/\sigma_2^2), \end{aligned}$$

where $\mu_i \in R$, $\underset{\sim}{1}_i = (1, \ldots, 1)' \in R^{n_i}$, $i = 1, 2$, $\sigma_1, \sigma_2 > 0$ and $x \in R^{n_1}$, $y \in R^{n_2}$. Here we consider the problem of testing H: $\sigma_1^2 = \sigma_2^2$ versus K: $\sigma_1^2 > \sigma_2^2$. Of course, if x and y are both normal with zero means, then the F-test based on $x'x/y'y$ is UMPI. We shall see that it is also UMPI in the above situation under a mild condition on q_1, q_2. This implies an optimality robustness of the F-test.

We assume for simplicity that μ_1 and μ_2 are known and set them equal to 0 without any loss of generality. The case when μ_1 and μ_2 are unknown is straightforward and left as an exercise (Exercise 8.3). Then the problem remains invariant under the transformation $x \to cx$, $y \to cy$, $\sigma_1 \to c\sigma_1$, $\sigma_2 \to c\sigma_2$ where $c \in \mathscr{G} = (0, \infty)$. As a left-invariant measure on $\mathscr{G}$, we take $\nu(dc) = dc/c$. Then the ratio R of the pdf's of a maximal invariant T under nonnull to null is given by $R = N/D$ where

$$\text{(8.14)} \qquad N \equiv N(\sigma_1, \sigma_2) = \int_{\mathscr{G}} \left(\prod_{i=1}^{2} \sigma_i^{-n_i}\right) q_1(c^2x'x/\sigma_1^2)q_2(c^2y'y/\sigma_2^2)c^{n-1}\, dc,$$

where $n = n_1 + n_2$ and $D = N(\sigma_1, \sigma_1)$. In (8.14), by transforming c into

$c(y'y)^{1/2}/\sigma_2$, N becomes

(8.15) $$N_\gamma = (y'y)^{-n/2}\gamma^{n_1/2}\int_0^\infty q_1(\gamma tc^2)q_2(c^2)c^{n-1}\,dc,$$

where $\gamma = \sigma_2^2/\sigma_1^2$ and $t = x'x/y'y$. Note that $D = N_1$. We now make the following assumption.

Assumption 8.3. $q_1, q_2 \in \mathcal{Q} = \{q\colon q(\beta u)/q(u)$ is nondecreasing in u for any $0 < \beta < 1\}$.

Theorem 8.3. *The test based on $t = x'x/y'y > k$ is UMPI for testing* H: $\sigma_1^2 = \sigma_2^2$ *versus* K: $\sigma_1^2 > \sigma_2^2$ *under the pdf's in* (8.13) *under the Assumption* 8.3. *But the null distribution of t depends on q_1 and q_2.*

Proof. Clearly t is a maximal invariant under $\mathcal{G}$. Under the null hypothesis, the pdf of t is directly shown to be (Exercise 8.4)

(8.16) $$dP_H^t/dt = at^{(n-2)/2}\int_0^\infty q_1(tc^2)q_2(c^2)c^{n-1}\,dc,$$

where a is a constant depending on n_1 and n_2 only, and P_H^t is the distribution of t under H: $\sigma_1^2 = \sigma_2^2$. Hence from $R = dP_K^t/dP_H^t = N_\gamma/N_1$ with N_γ in (8.15) and from (8.16), the pdf of t under (σ_1, σ_2) is given by

(8.17) $$h(t\,|\,\gamma)\,dt = R\,dP_H^t = at^{(n-2)/2}\gamma^{n_1/2}\left[\int_0^\infty q_1(\gamma tc^2)q_2(c^2)c^{n-1}\,dc\right]dt,$$

so $h(t\,|\,\gamma)/h(t\,|\,1) = R$. Further R in this sense is expressed as

(8.18) $$R = \gamma^{n_1/2}E[q_1(\gamma c^2)/q_1(c^2)],$$

where the expectation is taken with respect to c having the pdf,

(8.19) $$k_t(c^2) = q_1(c^2)q_2(c^2/t)c^{n-1}\bigg/\int_0^\infty q_1(c^2)q_2(c^2/t)c^{n-1}\,dc.$$

Here we show that $k_t(c^2)$ with $t > 0$ has a monotone likelihood ratio property in c^2. In fact, since $q_2 \in \mathcal{Q}$, $k_{t_2}(c^2)/k_{t_1}(c^2) \propto q_2(c^2/t_2)/q_2(c^2/t_1)$ is increasing in c^2 for $t_2 > t_1$. Therefore, since $q_1(\gamma c^2)/q_1(c^2)$ inside of the expectation of (8.18) is increasing in c^2 by assumption, its expectation is increasing in t (see Lehmann, 1959, p. 74). This implies that R in (8.18) is a nondecreasing function of t for $0 < \gamma < 1$. Consequently, by the Neyman–

Pearson lemma, the test based on $t > k$ is UMP. Since t does not depend on the q_i's, it is UMPI for all $q_1, q_2 \in \mathcal{Q}$. But the null distribution of t in (8.16) cannot be free from q_i's. This proves our result.

Remark 8.8. If $\mu_1 = \mu_2 = 0$ is not assumed, in other words, if μ_1 and μ_2 are unknown, the UMPI test statistic is defined as $(x - \bar{x}\underset{\sim}{1}_1)'(x - \bar{x}\underset{\sim}{1}_1)/(y - \bar{y}\underset{\sim}{1}_2)'(y - \bar{y}\underset{\sim}{1}_2)$ where $\bar{x} = x'\underset{\sim}{1}_1/n_1$, $\bar{y} = y'\underset{\sim}{1}_2/n_2$, and the nondecreasingness of $q(\beta u)/q(u)$ in Assumption 8.3 is replaced by that of $\bar{q}(\beta u)/\bar{q}(u)$, where $\bar{q}(u) = \int_{-\infty}^{\infty} q(u + x)\,dx$ (Exercise 8.5).

Remark 8.9. The null distribution of t is given by (8.17) with $\gamma = 1$, which is (8.16). Hence for specific q_1 and q_2, significance points of t are obtained from it.

Remark 8.10. If q is differentiable, a necessary and sufficient condition for $q(\beta u)/q(u)$ to be nondecreasing in u for $0 < \beta < 1$ is given by nonincreasingness of $uq'(u)/q(u)$ in u (Exercise 8.6). This condition is satisfied for a multivariate t-distribution and also for more general normal mixtures.

8.5. Test of Equality of Two Scale Parameters—Nonexponential Case

Let x: $n_1 \times 1$ and y: $n_2 \times 1$ be independent random vectors with pdf's (with respect to the Lebesgue measure)

$$f_{\mu_1,\sigma_1}(x) = \sigma_1^{-n_1} q_1\left[\sum_1^{n_1}(x_i - \mu_1)/\sigma_1\right],$$

(8.20)

$$f_{\mu_2,\sigma_2}(y) = \sigma_2^{-n_2} q_2\left[\sum_1^{n_2}(y_i - \mu_2)/\sigma_2\right].$$

$$x_i \geq \mu_1, \quad y_i \geq \mu_2, \quad \mu_1, \mu_2 \in R, \quad \sigma_1, \sigma_2 > 0.$$

We consider the problem of testing H: $\sigma_1 = \sigma_2$ versus K: $\sigma_1 > \sigma_2$ when μ_1 and μ_2 are unknown. The problem remains invariant under the group $\mathcal{G}$ of transformations $x_i \to ax_i + c_1$, $y_i \to ay_i + c_2$, $\mu_1 \to a\mu_1 + c_1$, $\mu_2 \to a\mu_2 + c_2$, $\sigma_1 \to a\sigma_1$, $\sigma_2 \to a\sigma_2$ for $-\infty < c_1, c_2 < \infty$ and $a > 0$. Let $x_{(1)} = \min x_i$, $y_{(1)} = \min y_i$, $t_1 = \Sigma_1^{n_1}(x_i - x_{(1)})$, $t_2 = \Sigma_1^{n_1}(y_i - y_{(1)})$. As a left-invariant measure ν on $\mathcal{G}$, we take $\nu(dg) = dc_1 dc_2 da/a$, where $g = (a, c_1, c_2)$. Then, by using Wijsman's theorem (Chapter 2), the ratio R of the pdf's of a maximal invariant T under nonnull to null is given by

$$(8.21)\qquad R = \frac{dP_{\mathrm{K}}^{T}}{dP_{\mathrm{H}}^{T}}$$

$$= \sigma_2^{-n_2}\int_A q_1\left[\frac{at_1 + n_1(ax_{(1)} + c_1 - \mu_1)}{\sigma_1}\right]$$

$$\times\, q_2\left[\frac{at_2 + n_2(ay_{(1)} + c_2 - \mu_2}{\sigma_2}\right]a^{n_1+n_2-1}\,da\,dc_1\,dc$$

$$\div\, \sigma_1^{-n_2}\int_A q_1\left[\frac{at_1 + n_1(ax_{(1)} + c_1 - \mu_1)}{\sigma_1}\right]$$

$$\times\, q_2\left[\frac{at_2 + n_2(ay_{(1)} + c_2 - \mu_2)}{\sigma_1}\right]a^{n_1+n_2-1}\,da\,dc_1\,dc$$

$$= \frac{\displaystyle\int_0^\infty \tilde{q}_1\left(\frac{at_1}{\sigma_1}\right)\tilde{q}_2\left(\frac{at_2}{\sigma_2}\right)\sigma_2^{-(n_2-1)}\,da}{\displaystyle\int_0^\infty q_1\left(\frac{at_1}{\sigma_1}\right)q_2\left(\frac{at_2}{\sigma_1}\right)\sigma_1^{-(n_2-1)}\,da}$$

$$= \delta^{-(n_2-1)}\cdot\frac{\displaystyle\int_0^\infty \tilde{q}_1(az\delta)\tilde{q}_2(a)\,da}{\displaystyle\int_0^\infty \tilde{q}_1(az)\tilde{q}_2(a)\,da}$$

$$= \delta^{-(n_2-1)}\frac{\displaystyle\int_0^\infty \tilde{q}_1(a\delta)\tilde{q}_2(az^{-1})\,da}{\displaystyle\int_0^\infty \tilde{q}_1(a)\tilde{q}_2(az^{-1})\,da},$$

where $A = \{a, c_1, c_2)$: $a > 0$, $c_1 > \mu_1 - ax_{(1)}$, $c_2 > \mu_2 - ay_{(1)}\}$, $\tilde{q}(x) = \int_0^\infty q(x + u)\,du$, $\delta = \sigma_2/\sigma_1$ and $z = t_1/t_2$. The last ratio of the integrals in (8.21) can be expressed as $E[\tilde{q}_1(a\delta)/\tilde{q}_2(a)]$, where E stands for expectation with respect to a having the pdf

$$(8.22)\qquad a \sim \tilde{q}_1(a)\tilde{q}_2(az^{-1})\bigg/\int_0^\infty \tilde{q}_1(a)\tilde{q}_2(az^{-1})\,da, \qquad a > 0.$$

We now make the following assumption.

Assumption 8.4. $\tilde{q}_1, \tilde{q}_2 \in \tilde{\mathcal{Q}} = \{\tilde{q}: \tilde{q}(ad)/\tilde{q}(\mathrm{a})$ is nondecreasing in a for $0 < d < 1\}$.

As in the previous sections, it then follows easily that the family (8.22) has MLR property in a, and hence R in (8.21) is nondecreasing in z for $\delta < 1$ (Lehmann, 1959, page 74). Thus we have proved the following result (Kariya and Sinha, 1987).

Theorem 8.4. *For testing* H: $\sigma_1 = \sigma_2$ *versus* K: $\sigma_1 > \sigma_2$ *in the model* (8.20), *the test with the critical region* $z > k$ *is UMPI whenever* $\tilde{q}_1, \tilde{q}_2 \in \tilde{\mathcal{Q}}$.

Remark 8.11. If either μ_1 or μ_2 or both are known, t_1 and t_2 are suitably redefined and Assumption 8.4 is modified accordingly.

Remark 8.12. The null pdf of z is given by

$$z^{-(n_2+3)/2} \int_0^\infty \tilde{q}_1(a)\tilde{q}_2(az^{-1})a^{(n_1+n_2)/2}\, da. \tag{8.23}$$

Hence for specific q_1 and q_2, significance points of z are obtained from (8.23).

Remark 8.13. If $\tilde{q}$ admits a derivative, $\tilde{\mathcal{Q}}$ is equivalent to

$$\tilde{\mathcal{Q}} = \{\tilde{q}\colon a\tilde{q}'(a)/\tilde{q}(a) \quad \text{is nonincreasing in} \quad a > 0\}.$$

An example of a $\tilde{q} \in \tilde{\mathcal{Q}}$ is provided by $\tilde{q}(u) \propto e^{-u^r}$, $r > 0$.

Exercises

8.1. Assume $q(u) > 0$ to be differentiable. Prove that $q((u-a)^2 + t)/q(u^2 + t)$ is increasing in u for all $a, t > 0$ is equivalent to $uq'(u^2 + t)/q(u^2 + t)$ is decreasing in u for all $t > 0$.

8.2. Let $q(u) > 0$ be differentiable. Prove that if $q(t + u - a)/q(t + u)$ is nondecreasing in u for $u > a$ and for all $t > 0$, then $uq'(t + u)/q(t + u)$ is nonincreasing in u for all $t \geq 0$. Show that the converse is also true.

8.3. Prove that if μ_1 and μ_2 in (8.13) are unknown, the UMPI test rejects $H\colon \sigma_1^2 = \sigma_2^2$ versus $K\colon \sigma_1^2 > \sigma_2^2$ for large values of $(x - \bar{x}\underset{\sim}{1}_1)'(x - \bar{x}\underset{\sim}{1}_1)/(y - \bar{y}\underset{\sim}{1}_2)'(y - \bar{y}\underset{\sim}{1}_2)$.

Hints: Note that here the appropriate group is

$$x \to cx + c_1\underset{\sim}{1}_1, \qquad y \to cy + c_2\underset{\sim}{1}_2, \quad c > 0, \quad c_1, c_2 \in R.$$

8.4. Prove that if μ_1 and μ_2 in (8.13) are 0, the pdf of $t = x'x/y'y$ is given by (8.16).

8.5. Under the setup of Exercise 8.3 above, show that the appropriate assumption analogous to Assumption 8.3 requires replacing $q(u)$ by $\bar{q}(u) = \int_{-\infty}^{\infty} q(u + x)\, dx$.

8.6. Assume $q(u) > 0$ to be differentiable. Prove that $q(\beta u)/q(u)$ is nondecreasing in u for $0 < \beta < 1$ is equivalent to $uq'(u)/q(u)$ is nonincreasing in u.

References

ANDERSON, T. W. (1948). "On the theory of testing serial correlation," *Skand, Aktuarie Tidskrt.* **31,** 88–116.

ANDERSON, T. W. (1958). *An Introduction to Multivariate Statistical Analysis.* Wiley, New York.

ANDERSON, T. W. (1971). *The Statistical Analysis of Time Series.* Wiley, New York.

ANDERSON, T. W. (1984). *An Introduction to Multivariate Statistical Analysis.* 2d ed. Wiley, New York.

ANDERSON, T. W., and DAS GUPTA, S. (1964). "Monotonicity of the power functions of some tests of independence between two sets of variates," *Ann. Math Statist.* **35,** 206–208.

ANDERSSON, S. A. (1978). *Invariant Measures.* Stanford Univ. Tech. Rep. No. 129.

ANDERSSON, S. A. (1982). "Distributions of maximal invariants using quotient measures," *Ann. Statist.* **10,** 955–961.

ANDERSSON, S. A., BRONS, H. K., and JENSEN, S. T. (1983). "Distribution of eigenvalues in multivariate statistical analysis," *Ann. Statist.* **11,** 392–415.

BANKEN, L. (1984). "On the reduction of the general MANOVA model" Tech. Rep. Universitat Trier, Trier.

BARNETT, V., and LEWIS, T. (1978). *Outliers in Statistical Data.* Wiley, New York.

BERBERIAN, S. K. (1965). *Measure and Integration.* Chelsea, New York.

BERGER, J. (1976). "Admissibility results for generalized Bayes estimators of coordinates of a location vector," *Ann. Statist.* **4,** 334–356.

BONDAR, J. V. (1976). "Borel cross-sections and maximal invariants," *Ann. Statist.* **4,** 866–877.

Box, G. E. P., and Watson, G. S. (1962). "Robustness to nonnormality of regression tests," *Biometrika* **49,** 93–106.

Bourbaki, N. (1963). *Elements de Mathématique,* **29;** Livre VI, Chap. 7. Hermann, Paris.

Bourbaki, N. (1966). *Elements of Mathematics;* General Topology, Part I. Addison-Wesley, Reading, Mass.

Chmielewski, M. A. (1980). "Invariant scale matrix hypothesis tests under elliptical symmetry," *J. Multivariate Anal.* **10,** 343–350.

Cohen, A., Rukhin, A., and Strawderman, W. E. (1986). "A characterization of the multivariate normal distribution and some remarks on linear estimators." Mimeo.

Das, R., and Sinha, B. K. (1986). "Detection of multivariate outliers with dispersion slippage in elliptically symmetric distributions," *Ann. Statist.* **14,** 1619–1624.

Das Gupta, S., and Perlman, M. (1972). "On the power of the noncentral F- and Wilks' U-tests: Choosing variates for increasing the power of Hotelling's T^2-test," *J. Amer. Statist. Assoc.* **69,** 174–180.

Dawid, A. P. (1977). "Spherical matrix distributions and a multivariate model," *J. Roy. Statist. Soc. B,* **39,** 254–261.

Dempster, A. P. (1969). *Elements of Continuous Multivariate Analysis.* Addison-Wesley, Reading, Mass.

Durbin, J., and Watson, G. S. (1950). "Testing for serial correlation in least squares regression I," *Biometrika* **37,** 409–428.

Durbin, J., and Watson, G. S. (1951). Testing for serial correlation in least squares regression II," *Biometrika* **38,** 159–178.

Durbin, J., and Watson, G. S. (1971). "Testing for serial correlation in least squares regression III," *Biometrika* **58,** 1–19.

Eaton, M. L. (1981). "On the projections of isotropic distributions," *Ann. Statist.* **9,** 391–400.

Eaton, M. L. (1983). *Multivariate Statistics: A Vectorspace Approach.* Wiley, New York.

Eaton, M. L. (1988). "Group invariance in statistics and its applications," SIAM Lecture Notes.

Eaton, M. L., and Kariya, T. (1984). "A condition for null robustness," *J. Multivariate Anal.* **14,** 155–168.

Farrell, R. H. (1962). "Representation of invariant measures," *Illinois J. Math* **7,** 447–467.

Farrell, R. (1976). *Techniques of Multivariate Calculation.* Springer-Verlag, New York.

Farrell, R. H. (1985). *Multivariate Calculation.* Springer-Verlag, New York.

Fearn, T. (1975). "A Bayesian approach to growth curves," *Biometrika* **62,** 89–100.

Feller, W. (1966). *An Introduction to Probability and Its Applications,* **2.** Wiley, New York.

FERGUSON, T. S. (1961). "On the rejection of outliers," *Proc. Fourth Berk. Symp. Math. Statist. Prob. I,* 253–287.

FERGUSON, T. S. (1967). *Mathematical Statistics—A Decision Theoretic Approach.* Academic Press, New York.

FRASER, D. A. S., and NG, K. W. (1980). *Multivariate Regression Analysis with Spherical Error.* Krishnaiah, P. R. (ed.). Multivariate Analysis-V. North-Holland, New York, 369–386.

FUJIKOSHI, Y. (1973). "Monotonicity of the power functions of some tests in general MANOVA models," *Ann. Statist.* **1,** 388–391.

GEISSER, S. (1970). "Bayesian analysis of growth curves," *Sankhya A* **32,** 53–64.

GEISSER, S. (1980). "Growth curve analysis." In *Handbook of Statistics, 1* (ed. Krishnaiah, P. R.). North-Holland, New York, 89–115.

GIRI, N. C. (1968). "Locally and asymptotic minimax tests of a multivariate problem," *Ann. Math. Statist.* **39,** 171–178.

GIRI, N. C. (1977). *Multivariate Statistical Inference.* Academic Press, New York.

GIRI, N., and KIEFER, J. (1964). "Local and asymptotic minimax properties of a normal multivariate testing problem," *Ann. Math. Statist.* **35,** 21–35.

GLESER, L. J., and OLKIN, I. (1970). "Linear models in multivariate analysis." *Essays in Probability and Statistics.* 267–292. Univ. of North Carolina Press, Chapel Hill, North Carolina, Edited by Bose, R. C., et al.

GNANADESIKAN, R. (1977). *Methods for Statistical Data Analysis of Multivariate Observations.* Wiley, New York.

HALL, W. J., WIJSMAN, R. A., and GHOSH, J. K. (1965). "The relationship between sufficiency and invariance with applications in sequential analysis," *Ann. Math. Statist.* **36,** 575–614.

HAWKINS, D. M. (1980). *Identification of Outliers.* Chapman and Hall, New York.

HOOPER, P. M. (1982). "Invariant confidence sets with smallest expected measure," *Ann. Statist.* **10,** 1283–1294.

HOOPER, P. M. (1983). "Simultaneous interval estimation in the general multivariate analysis of variance model," *Ann. Statist.* **11,** 666–673.

JAMES, A. T. (1964). "Distributions of matrix variates and latent roots derived from normal samples," *Ann. Math Statist.* **35,** 475–501.

JENSEN, D. R. (1979). "Linear models without moments," *Biometrika* **66,** 611–617.

JENSEN, D. R., and GOOD, I. J. (1981). "Invariant distributions associated with matrix laws under structural symmetry," *J. Roy. Statist. Soc. B* **43,** 327–332.

JOHN, S. (1971). "Some optimal multivariate tests," *Biometrika* **58,** 123–127.

JOHNSON, N. L., and KOTZ, S. (1972). *Distributions in Statistics: Continuous Multivariate Distributions.* Wiley, New York.

KARIYA, T. (1977). "A robustness property of the tests for serial correlation," *Ann. Statist.* **5,** 1212–1220.

KARIYA, T. (1978). "The general MANOVA problem," *Ann. Statist.* **6,** 200–214.

KARIYA, T. (1980). "Locally robust tests for serial correlation in least squares regression," *Ann. Statist.* **8,** 1065–1070.

KARIYA, T. (1981a). "Robustness of multivariate tests," *Ann. Statist.* **9,** 1267–1275.

KARIYA, T. (1981b). "A robustness property of Hotelling's T^2-problem," *Ann. Statist.* **9,** 210–214.

KARIYA, T. (1985). *Testing in the Multivariate General Linear Model.* Kinokuniya, New York.

KARIYA, T., and EATON, M. L. (1977). "Robust tests for spherical symmetry," *Ann. Statist.* **5,** 206–215.

KARIYA, T., and KANAZAWA, M. (1978). "A locally best invariant test for the equality of means in the presence of covariates," *J. Multivariate Anal.* **8,** 134–140.

KARIYA, T., and SINHA, B. K. (1985). "Nonnull and optimality robustness of some tests," *Ann. Statist.* **13,** 1182–1197.

KARIYA, T., and SINHA, B. K. (1987). "Optimality robustness of tests in two population problems" *Journal of Statistical Planning and Inference* **15,** 167–176.

KARIYA, T., SINHA, B. K., and GIRI, N. C. (1987). "Robustness of t-test," *J. Japan Statist. Soc.* **17,** 165–173.

KELKER, D. (1970). "Distribution theory of spherical distribution and a location scale parameter generalization," *Sankhya A* **43,** 419–430.

KHATRI, C. G. (1966). "A note on MANOVA model applied to problems in growth curves," *Ann. Inst. Statist. Math.* **18,** 75–86.

KIEFER, J. and SCHWARTZ, R. (1965). "Admissible Bayes character of T^2-, R^2-, and other fully invariant tests for classical multivariate normal problems," *Ann. Math. Statist.* **36,** 747–770.

KOEHN, U. (1970). "Global cross sections and the densities of maximal invariants" *Ann. Math. Statist.* **41,** 2045–2056.

KRISHNAIAH, P. R. (1966). "Simultaneous test procedures for general MANOVA models," *Multivariate Analysis* **II,** 121–143.

KRISHNAIAH, P. R. (1979). "Some recent developments on real multivariate distributions." In *Developments in Statistics,* (ed. P. R. Krishnaiah) **1.** Academic Press, New York, 135–169.

LEE, J. C., and GEISSER, S. (1972). "Growth curve prediction," *Sankhya A* **34,** 393–412.

LEE, J. C., and GEISSER, S. (1975). "Applications of growth curve prediction," *Sankhya A* **37,** 239–256.

LEHMANN, E. L., and STEIN, C. (1948). "Most powerful tests of composite hypotheses," *Ann. Math. Statist.* **19,** 495–516.

LEHMANN, E. L., and STEIN, C. (1949). "On the theory of some nonparametric hypotheses," *Ann. Math. Statist.* **20,** 28–45.

LEHMANN, E. L. (1959). *Testing Statistical Hypotheses.* Wiley, New York.

MARDEN, J. I. (1983). "Admissibility of invariant tests in the general multivariate analysis of variance problem," *Ann. Statist.* **11,** 1086–1099.

MARDEN, J., and PERLMAN, M. D. (1980). "Invariant tests for means with covariates," *Ann. Statist.* **8,** 25–63.

MARDIA, K. V. (1970). "Measures of multivariate skewness and kurtosis with applications," *Biometrika* **57,** 519–530.

MUIRHEAD, ROBB (1982). *Aspects of Multivariate Statistical Theory.* Wiley, New York.

NABEYA, S., and KARIYA, T. (1986). "Transformations preserving normality and Wishart-ness," *J. Multivariate Anal.* **20,** 251–264.

NACHBIN, L. (1965). *The Haar Integral.* Van Nostrand, New York.

NIMMO-SMITH, I. (1979). "Linear regressions and sphericity," *Biometrika* **66,** 390–392.

OLKIN, I., and RUBIN, H. (1964). "Multivariate beta distributions and independence properties of the Wishart distribution," *Ann. Math. Statist.* **35,** 261–269.

PALAIS, R. S. (1961). "On the existence of slices for actions of noncompact Lie groups," *Ann. Math.* **73,** 295–323.

PERLMAN, M. D., and OLKIN, I. (1980). "Unbiasedness of invariant tests for MANOVA and other multivariate problems," *Ann. Statist.* **8,** 1326–1341.

PONTRYAGIN, L. (1966). *Topological Groups,* 2d ed. Translated from the Russian, Gordon and Breach, New York.

POTTHOFF, R. F., and ROY, S. N. (1964). "A generalized multivariate analysis of variance model useful especially for growth curve models," *Biometrika* **51,** 313–326.

PRESS, S. J. (1969). "On serial correlation," *Ann. Math. Statist.* **40,** 186–196.

RAO, C. R. (1965). "The theory of least squares when the parameters are stochastic and its application to the analysis of growth curves," *Biometrika* **52,** 447–458.

RAO, C. R. (1967). "Least squares theory using an estimated dispersion matrix and its application to measurement of signals," *Proc. Fifth Berk. Symp. Math. Statist. Prob.* **1,** 355–372.

VON ROSEN, K. (1985). *Multivariate Linear Normal Model with Special References to the Growth Curve Model.* Holms gards tryckeri, Edsbruk.

ROYDEN, H. L. (1968). *Real Analysis.* MacMillan, New York.

SCHOENBERG, I. J. (1938). "Metric spaces and completely monotone functions," *Ann. Math.* **39,** 811–841.

SCHWAGER, S. J., and MARGOLIN, B. H. (1982). Detection of multivariate normal outliers," *Ann. Statist.* **10,** 943–954.

SCHWARTZ, R. (1967a). "Local minimax tests," *Ann. Math. Statist.* **38,** 340–360.

SCHWARTZ, R. (1967b). "Admissible tests in multivariate analysis of variance" *Ann. Math. Statist.* **38,** 698–710.

SEGAL, I. E., and KUNZE, R. A. (1968). *Integrals and Operators.* Tata McGraw-Hill, Bombay.

SINHA, B. K. (1984). "Detection of multivariate outliers in elliptically symmetric distributions," *Ann. Statist.* **12,** 1558–1565.

SINHA, S. K. (1986). *Reliability and Life Testing.* Halsted Press/Wiley Eastern Ltd. New Delhi.

SIOTANI, M., HAYAKAWA, T., and FUJIKOSHI, Y. (1985). *Modern Multivariate Statistical Analysis: A Graduate Course and Handbook.* American Sciences Press, Columbus.

STEIN, C. (1966). "Notes on multivariate analysis." Stanford Univ. (Notes recorded by M. L. Eaton).

SUGIURA, N. (1972). "Locally best invariant test for sphericity and the limiting distributions," *Ann. Math. Statist.* **43,** 1312–1316.

SUGIURA, N., and NAGAO, H. (1968). "Unbiasedness of some test criteria for the equality of one or two covariance matrices," *Ann. Math. Statist.* **39,** 1686–1692.

WALD, A. (1950). *Statistical Decision Functions.* Wiley, New York.

WARE J. H., and BOWDEN, R. E. (1977). "Circadian rhythm analysis when output is collected at intervals," *Biometrics* **33,** 566–571.

WIJSMAN, R. A. (1959). "Regular best asymptotically normal estimators," *Ann. Math. Statist.* **30,** 185–195; correction: 1296–1298.

WIJSMAN, R. A. (1967). "Cross-sections of orbits and their application to densities of maximal invariants," *Proc. Fifth Berk. Symp. Math. Statist. Prob.* **1,** 389–400. Univ. of California, Berkeley.

WIJSMAN, R. A. (1969). "General proof of termination with probability one of invariant sequential probability ratio tests based on multivariate observations," *Ann. Math. Statist.* **38,** 8–24.

WIJSMAN, R. A. (1978). "Distribution of maximal invariants, using global and local cross-sections." Preprint, Univ. of Illinois at Urbana–Champaign.

WIJSMAN, R. A. (1985). "Proper action in steps, with application to density ratios of maximal invariants," *Ann. Statist.* **13,** 395–402.

WIJSMAN, R. A. (1986). "Global cross sections as a tool for factorization of measures and distribution of maximal invariants," *Sankhya A* **48,** 1–42.

WILKINSON, G. N. (1958). "Estimation of the missing values for the analysis of incomplete data," *Biometrics* **14,** 257–286.

WILKS, S. S. (1962). *Mathematical Statistics.* Wiley, New York.

ZERBE, G. O., and JONES, R. H. (1980). "On application of growth curve techniques to time series data," *J. Amer. Statist. Assoc.* **75,** 507–509.

Author Index

H

J

K

L

M

N

O

P

R

S

W

Z

Subject Index

STATISTICAL MODELING AND DECISION SCIENCE

INGRAM OLKIN AND GERALD LIEBERMAN, EDITORS

Samuel Eilon, *The Art of Reckoning: Analysis of Performance Criteria*

Enrique Castillo, *Extreme Value Theory in Engineering*

Joseph L. Gastwirth, *Statistical Reasoning in Law and Public Policy: Volume 1, Statistical Concepts and Issues of Fairness; Volume 2, Tort Law, Evidence, and Health*

Takeaki Kariya and Bimal K. Sinha, *Robustness of Statistical Tests*